Advanced Structured Materials

Volume 107

Common engineering materials reach in many applications their limits and new developments are required to fulfil increasing demands on engineering materials. The performance of materials can be increased by combining different materials to achieve better properties than a single constituent or by shaping the material or constituents in a specific structure. The interaction between material and structure may arise on different length scales, such as micro-, meso- or macroscale, and offers possible applications in quite diverse fields.

This book series addresses the fundamental relationship between materials and their structure on the overall properties (e.g. mechanical, thermal, chemical or magnetic etc.) and applications.

The topics of *Advanced Structured Materials* include but are not limited to

- classical fibre-reinforced composites (e.g. glass, carbon or Aramid reinforced plastics)
- metal matrix composites (MMCs)
- micro porous composites
- micro channel materials
- multilayered materials
- cellular materials (e.g., metallic or polymer foams, sponges, hollow sphere structures)
- porous materials
- truss structures
- nanocomposite materials
- biomaterials
- nanoporous metals
- concrete
- coated materials
- smart materials

Advanced Structured Materials is indexed in Google Scholar and Scopus.

More information about this series at http://www.springer.com/series/8611

V. E. Gromov · S. V. Konovalov · Yu. F. Ivanov · K. A. Osintsev

Structure and Properties of High-Entropy Alloys

Springer

V. E. Gromov
Siberian State Industrial University
Novokuznetsk, Russia

S. V. Konovalov
Samara National Research University
Samara, Russia

Yu. F. Ivanov
Institute of High Current Electronics SB
Tomsk, Russia

K. A. Osintsev
Samara National Research University
Samara, Russia

ISSN 1869-8433 ISSN 1869-8441 (electronic)
Advanced Structured Materials
ISBN 978-3-030-78363-1 ISBN 978-3-030-78364-8 (eBook)
https://doi.org/10.1007/978-3-030-78364-8

This Springer imprint is published by the registered company Springer Nature Switzerland AG
The registered company address is: Gewerbestrasse 11, 6330 Cham, Switzerland

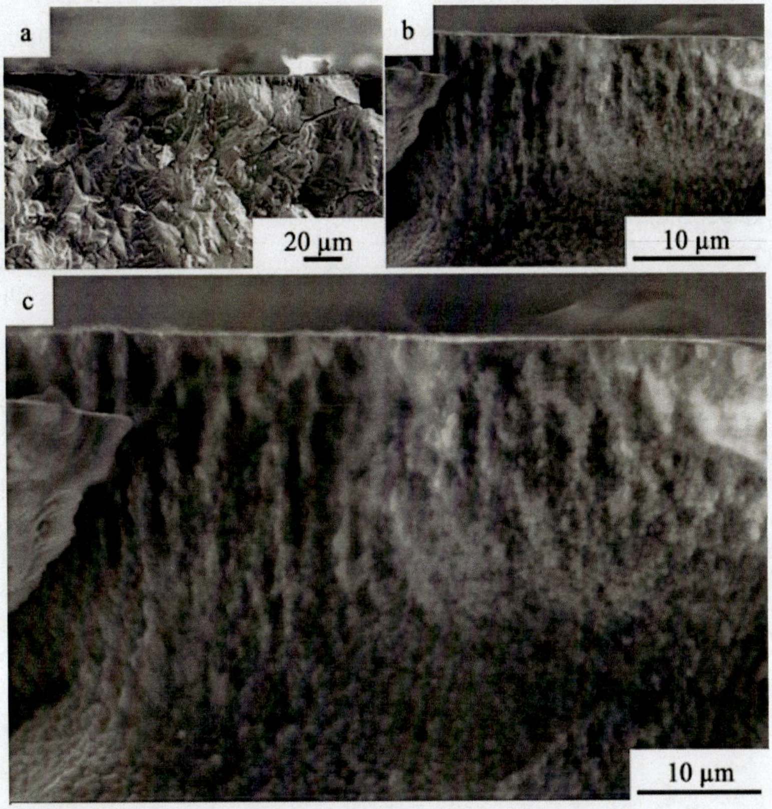

Fig. 6.24 SEM image of HEA fracture structure processed by pulsed electron beam with energy density of 30 J/cm^2 (200 μs, 3 pulses)

Table 6.5 Comparative data on layer thickness and surface substructure size depending on energy density of electron beam

Characteristics of columnar structure	Energy density electron beam, J/cm^2				
	10	15	20	25	30
H, μm	0, 8–1	9–10	13	13–15	20
h, μm	Not revealed	0.15–0.3	0.3	1.0	2.5–3.0

The investigations into the elemental composition of samples irradiated by pulsed electron beam are performed by methods of micro-X-ray spectral analysis of transverse metallographic sections. It is found that electron beam processing results in a more homogeneous distribution of elements in the processed layer. So, the enrichment in aluminium and nickel of grain volume being observed in the unprocessed alloy as well as the enrichment in iron and chromium (Fig. 6.4) of grain boundaries are not observed in the modified alloy. Mapping of sample cross-section depending on the distance from surface showed that thickness of homogenized layer reaches

4 μm. With a larger increase in distance from irradiation surface a distribution of elements characteristic of the unprocessed sample is observed.

By the methods of energy dispersion spectral analysis, it was detected that on average the processed layer has the following elemental composition: 39.05 at. % Al, 4.88 at. % Co, 7.92 at. % Cr, 15.9 at. % Fe, 32.25 at. % Ni. The data testify to the increase in aluminium content in the layer and the corresponding decrease in iron, nickel, cobalt and chromium.

The sample irradiation by pulsed electron beam leads to no change in material phase composition as evidenced by the similarity of diffraction patterns illustrated in Fig. 6.25.

The displacement of HEA diffraction lines to the site of large angles is observed after irradiation that is indicative of the possible change in alloy crystal lattice parameter. In fact, the results depicted in Fig. 6.26 show that alloy crystal lattice parameter after irradiation depends on the energy density of electron beam in a non-monotonous way and is characterized by essentially smaller value relative to the initial material. Simultaneously with the change in crystal lattice parameter the irradiation leads to the decrease in microdistortions of the alloy crystal lattice and the increase in sizes of coherent scattering regions.

The mechanical properties of HEA samples irradiated by pulsed electron beam were characterized by hardness. Microhardness measurement was done both on samples surface processed by electron beam and in transverse cross-section.

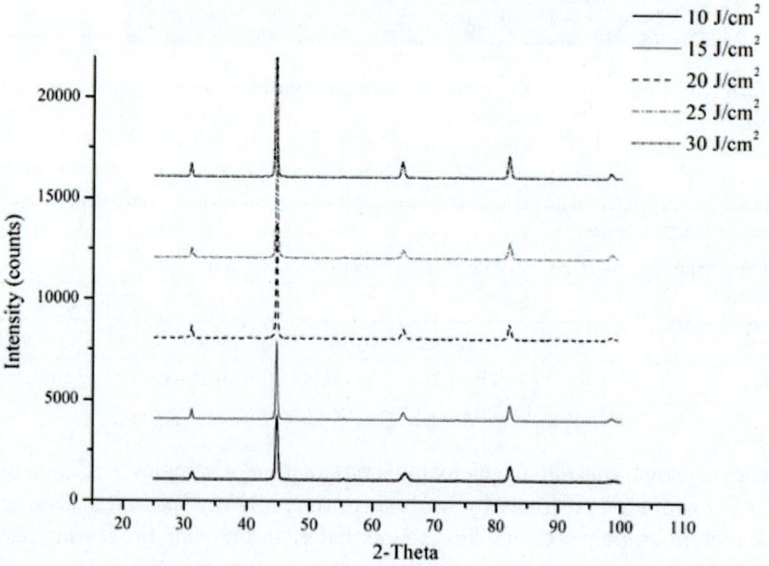

Fig. 6.25 HEA diffraction pattern obtained from surface processed by electron beams with different energy density

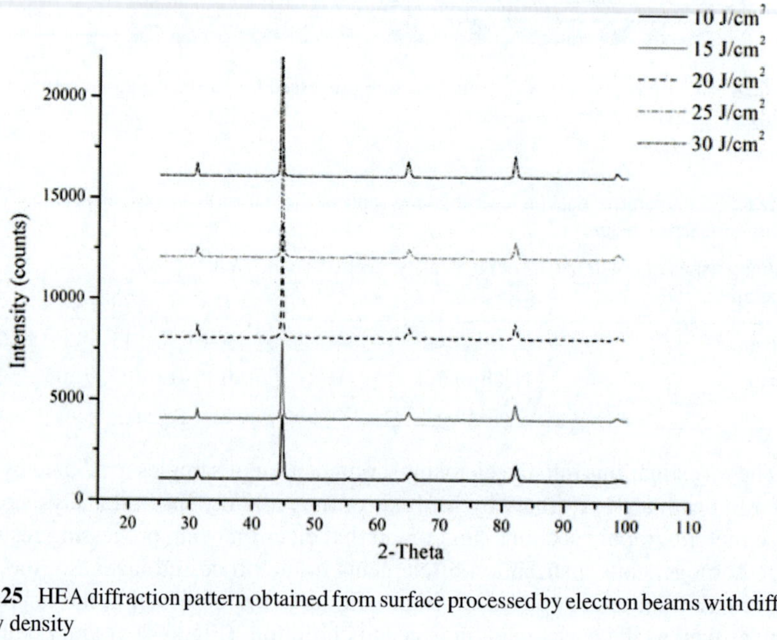

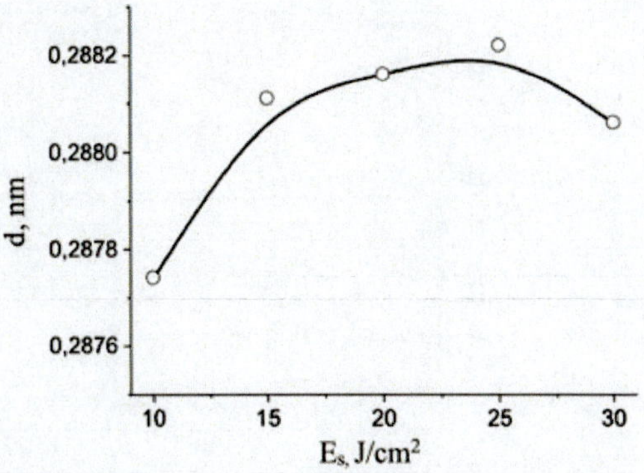

Fig. 6.26 A plot of HEA crystal lattice parameter as a function of the energy density of electron beams (200 μs, 3 puls.). The alloy crystal lattice parameter prior to irradiation (initial state) amounts to 0.28914 nm

It is detected that electron beam processing of alloy sample surface at different energy densities of electron beam resulted in the decrease in surface layer microhardness independent of processing parameters. The values decreased on average by 100 HV in the processed layer relative to material bulk. The least microhardness value was observed in processing mode with the energy density of electron beam of 10 J/cm^2, it amounted to 368 ± 1 HV on the surface. The largest microhardness value of the processed surface of 403 ± 6 HV was detected in the material processed with the energy density of an electron beam of 25 J/cm^2. It is established that electron beam processing results in a change in surface layer microhardness to the depth up to 90 μm.

The results of investigations into nanohardness and elastic modulus of processed samples revealed a correlation with data on change in microhardness, namely, nanohardness and Young's modulus of surface layer decreased 28–30% on the average. The fact testifies that electron beam processing leads to the relaxation of internal stress fields formed in the initial material on its manufacturing.

The performed tribological tests of samples in the initial state and after electron beam processing in different modes showed that electron beam processing affects slightly the friction coefficient and wear rate. The wear rate value of unprocessed sample amounts to $1.4 \cdot 10^{-4}$ mm^3/N·m, and in samples processed, by electron beam the values vary from 1.4 to $2.5 \cdot 10^{-4}$ mm^3/N·m. The friction coefficient amounts to 0.65 for initial samples and 0.63–0.67 for modified ones (Table 6.6).

Table 6.6 Results of tribological tests

Energy density of electron beam, J/cm^2	Wear rate, 10^{-4} mm^3/N·m	Friction coefficient
Initial sample	1.4	0.65
10	1.9	0.65
15	2.3	0.67
20	1.4	0.63
25	2.5	0.65
30	1.9	0.63

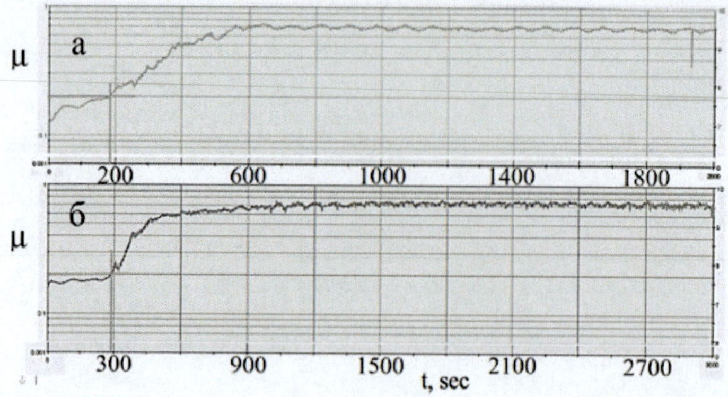

Fig. 6.27 Friction coefficient value versus duration of HEA samples tests for wear resistance curves. Lines designate test time at which friction coefficient value is less than 0.2

As a result of tribological tests, it is determined that the value of HEA friction coefficient at the initial stage of tests depends substantially on energy density of the electron beam. The test results presented in Fig. 6.27 show that at the initial stage of tests (up to 300 s) the friction coefficient of irradiated samples is less than 0.2. On increasing the test time, the friction coefficient increases and comes to a plateau corresponding to the friction coefficient value of the initial material ($\mu = 0.7$). It may be suggested that the change in friction coefficient is related to two moments. First, the presence of a modified layer after irradiation whose friction coefficient is small; second, the failure of the layer and participation of base material in wear.

If we take $\mu = 0.2$ as a benchmark, then it turns out that the friction coefficient depends nonmonotonously on energy density of electron beam in this test time interval. The best tribological properties are shown by the sample irradiated at energy density of electron beam of 20 J/cm^2, for which time interval reaches 300 s (Table 6.7).

Irradiation of high-entropy alloy by intense pulsed electron beam led to an increase in strength and plasticity of the material. So, ultimate compression strength increased 1.1–1.6 times (Fig. 6.28). The largest ultimate compression strength value of 2179 MPa was obtained in the alloy processed by electron beam with energy density of 30 J/cm^2. The conditional ultimate compression strength, in this case, amounted to 522 MPa, and Young's modulus amounted to 257 GPa (Table 6.8).

Characteristic image of surface being formed in HEA failure conditions of compression is shown in Fig. 6.29.

Table 6.7 Time of HEA-2 sample tests for friction versus energy density of electron beam at which friction coefficient is less than 0.2

Energy density of electron beam, J/cm^2	0	10	15	20	25	30
Test time, sec ($\mu = 0{,}2$)	180	240	256	292	143	128

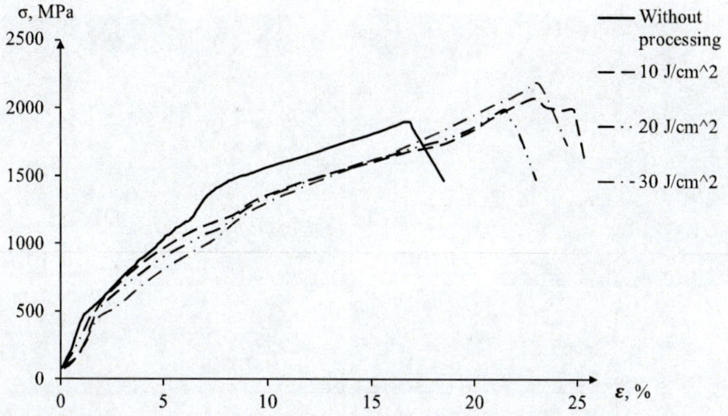

Fig. 6.28 Stress–strain diagrams of HEA samples in initial state and processed by electron beams

Table 6.8 Mechanical properties of HEA sample before and after electron beam processing revealed in uniaxial compression tests

	$\sigma_{0,2}$, MPa	σ_B, MPa	E, GPa	ε, %
Initial state	512–523	1361–1899	273–372	9–18
10 J/cm^2	522	2078	310	25
20 J/cm^2	474	2000	279	23
30 J/cm^2	522	2179	257	25

The investigations carried out revealed a failure structure similar to that being formed in HEA sample failure in conditions of uniaxial tension (see Figs. 6.17 and 6.18). Namely, in the bulk of grains a grooved pattern is present that is steps between different local cleavage facets of the same common plane (Fig. 6.29). The source of local failure is particles of the second phase located in the junction of grains (the site of particle location is indicated by arrows in Fig. 6.30). In Fig. 6.30 b the arrow indicates micropores which will also contribute to alloy embrittlement.

6.5 Conclusion

By means of the technology of wire-arc additive manufacturing the samples of high-entropy alloy (HEA) have been fabricated. A multi-component wire consisting of three strands of different compositions was used as the initial material. The obtained composition has the following elemental composition (wt. %): 15.64 Al, 7.78 Co, 8.87 Cr, 22.31 Fe, 44.57 Ni and corresponds to the non-equimolar high-entropy alloy of Al-Co-Cr-Fe-Ni system. It has been stated by the methods of optical and scanning electron microscopy of the transverse etched metallurgical section that deposited layers have a dendritic structure. Sizes of alloy grains vary in the limits from 4 to

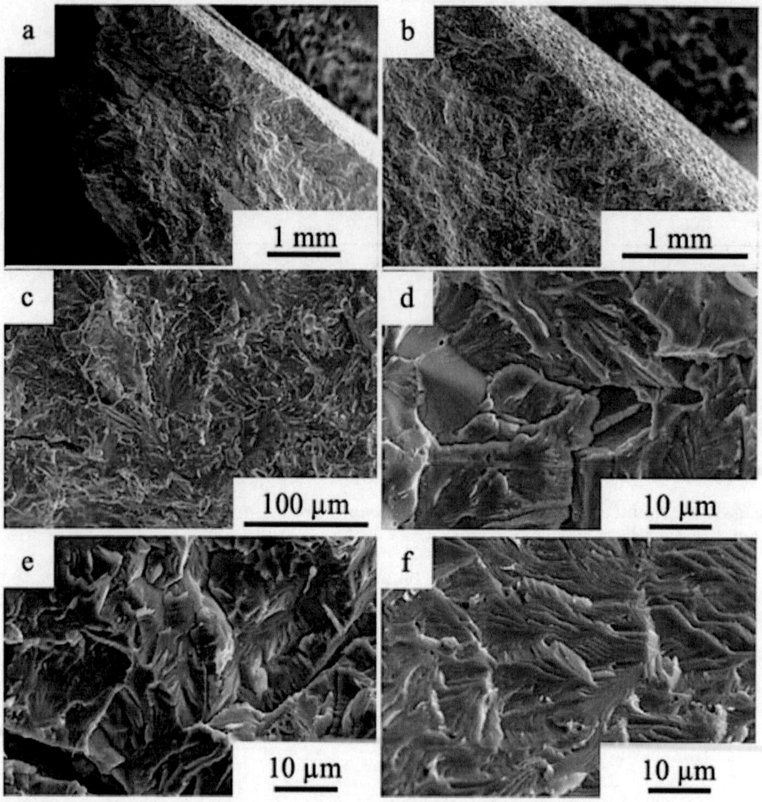

Fig. 6.29 Structure of failure surface being formed in HEA sample compression. Scanning electron microscopy

15 μm. Along boundaries and in the bulk of grains the second phase inclusions have been revealed. It has been determined by the methods of mapping that near-boundary bulks of the alloy (bulks located along grain boundaries) are enriched in chromium and iron atoms, the bulk of grains is enriched in nickel and aluminium atoms, cobalt is quasi-homogeneously distributed in the alloy. The melting temperature of obtained high-entropy alloy detected by the methods of thermogravimetric analysis is equal to 1495.18° C. By means of X-Ray phase analysis of the areas located at a distance of 15, 35 and 55 mm from the substrate the homogeneity of phase composition of the manufactured high-entropy alloy has been revealed. The crystal lattice parameter of this phase is 0.28914 nm. Taking into account the presence of second-phase inclusions in the alloy the results of X-ray phase analysis may be interpreted within the frameworks of material multi-phase composition with the following relation of phases: 36.56% AlNi, 27.02% CrFe, 36.42% Al_2FeCo. The presence of inclusions of spherical shape has been revealed by the methods of transmission electron microscopy in the alloy. By the mapping methods, it has been established that inclusions of extended shape are enriched in atoms of chromium, iron, and oxygen and may be carbides.

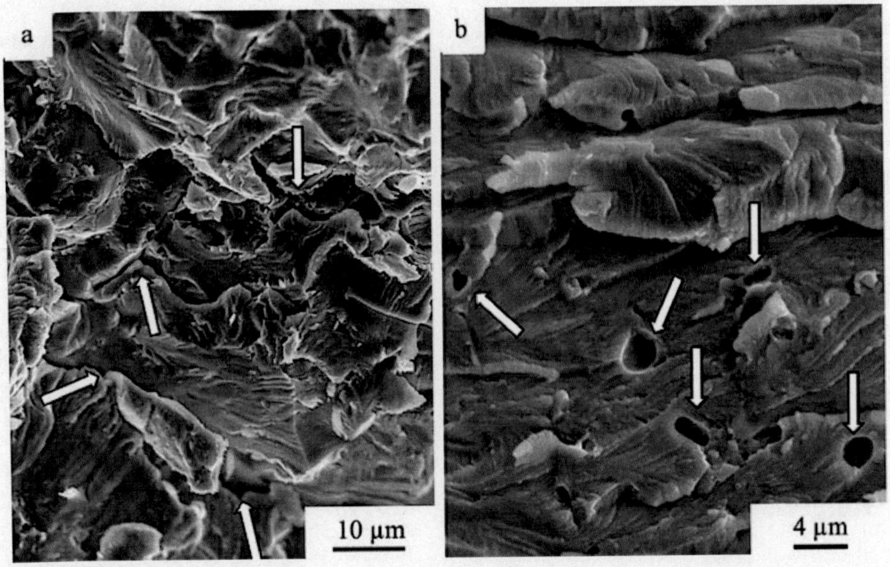

Fig. 6.30 Electron microscopic image of typical structural elements of HEA failure surface subjected to uniaxial compression strain. Arrows **a** designate second phase particles being a source of sample local failure. Arrows **b** indicate micropores

It has been established that average microhardness values of samples obtained by mode No. 1 amounts to 474 ± 18 HV, by mode No. 2—465 ± 3 HV, by mode No. 3—474 ± 8 HV. The elastic modulus determined by the nanoindentation method amounts to 304 ± 15 GPa. As a result of tribological tests, it has been identified that HEA wears coefficient amounts to 1.4×10^{-4} m^3/N·m and friction coefficient amounts to 0.65.

The results of the tests by the compression method have shown that the ultimate strength of the material manufactured by mode No. 1 is within the range from 652 to 856 MPa, by mode No. 2—from 1361 to 1899 MPa, and by mode No. 3—from 556 to 1390 MPa. The Young's modulus of samples manufactured by the modes has values of 165–212 GPa (mode No. 1), 273–372 GPa (mode No. 2), 192–270 GPa (mode No. 3).

The irradiation of the HEA sample surface by the pulsed electron beam in the mode of surface layer melting has been carried out. The fragmentation of sample surface with of microcracks' network has been determined. It has been established that HEA irradiation by the pulsed electron beam is accompanied by the release of grain boundaries from the second phase precipitations that is indicative of material homogenisation. In this case, the phase composition of the alloy is not changed. It has been shown that high-velocity crystallisation of molten surface layer of HEA samples is accompanied by the formation of columnar structure having a submicro-nanocrystalline structure. It has been established by the methods of micro-X-ray spectral analysis that the elemental composition of alloy surface layer is practically independent of the energy density of electron beam and within measurement error limits it corresponds to the elemental composition of initial material. It has been determined that electron beam processing results in decrease on microhardness of alloy surface layer up to 90 μm thick that may be caused by relaxation of internal field stresses formed in the initial material in manufacturing. It has been shown that HEA processing by pulsed electron beam affects slightly the friction coefficient and material wear rate. It has been revealed that irradiation of high-entropy alloy by intense pulsed electron beam results in the increase in strength and plasticity of the material. The ultimate compression strength increases 1.1–1.6 times. The largest value of ultimate compression strength of 2179 MPa was obtained in the alloy processed by electron beam with the energy density of 30 J/cm^2. The conditional compressive yield point, in this case, amounted to 522 MPa, and Young's modulus amounted to 257 GPa.

References

Chen, Q., Zhou, K., Jiang, L., Lu, Y., Li, T.: Arab. J. Sci. Eng. **40**, 3657 (2015)
Jiao, Z.M., Ma, S.G., Yuan, G.Z., Wang, Z.H., Yang, H.J., Qiao, J.W.: J. Mater. Eng. Perform. **24**, 3077 (2015)
Guo, J., Huang, X., Huang, W.: J. Mater. Eng. Perform. **26**, 3071 (2017)
Alagarsamy, K., Fortier, A., Komarasamy, M., Kumar, N., Mohammad, A., Banerjee, S., Han, H.C., Mishra, R.S.: Cardiovasc. Eng. Technol. **7**, 448 (2016)
Cao, L., Wang, X., Wang, Y., Zhang, L., Yang, Y., Liu, F., Cui, Y.: Appl. Phys. A Mater. Sci. Process. **125**, 1 (2019)
Chumak, I., Richter, K.W., Ipser, H.: J. Phase Equilibria Diffus. **29**, 300 (2008)
Ivanov, Y.F., Gromov, V.E., Konovalov, S.V.: Arab. J. Sci. Eng. **34**, 233 (2009)
Kang, M., Lim, K.R., Won, J.W., Na, Y.S.: J. Alloys Compd. **769**, 808 (2018)
Lin, C.-M., and Tsai, H.-L.: Intermetallics **19**, 288 (2011)
Qin, G., Xue, W., Fan, C., Chen, R., Wang, L., Su, Y., Ding, H., Guo, J.: Mater. Sci. Eng. A **710**, 200 (2018)
Sun, Y., Wu, C., Peng, H., Liu, Y., Wang, J., Su, X.: J. Phase Equilibria Diffus. **40**, 706 (2019)
Takeuchi, A., Inoue, A.: J. Postgrad. Med. Inst. **30**, 80 (2016)
Tian, Q., Zhang, G., Yin, K., Wang, W., Cheng, W., Wang, Y.: Mater. Charact. **151**, 302 (2019)
Wang, F.J., Zhang, Y., Chen, G.L.: J. Alloys Compd. **478**, 321 (2009)
Zhang, A., Han, J., Meng, J., Su, B., Li, P.: Mater. Lett. **181**, 82 (2016)
Zhang, L., Zhou, D., Li, B.: Mater. Lett. **216**, 252 (2018)

Introduction

One of the fundamental and practically oriented tasks of solid-state physics and physical materials science is the development of physical bases of creating new metallurgical materials with a complex of necessary physico-mechanical and operational properties and technologies of their production. As it is known, mechanical properties of materials depend essentially on their chemical composition and characteristic features of the structural state, such as type, grain size, form of their boundaries, grade, quantity and distribution in size and volume of excessive phases, density and type of dislocation substructure, etc.

Considerable improvement of complex of practical design, functional and technological parameters of alloys and intermetallics was connected with additional micro- and macro-alloying (with the third, fourth, fifth, sixth elements), development of special strengthening and plastifying technologies of both synthesis and subsequent treatment of poly- and monocrystals, modification of their micro- and submicrocrystalline structures. At the beginning of the twenty-first century the papers on creating and complex studying the new so-called high-entropy polymetallic alloys including 5–6 or more principal elements appeared.

In papers (Tong et al. 2005a, b; Chen et al. 2006; Li and Zhang 2009; Tsai et al. 2010; Braic et al. 2010; Huang and Yeh 2009; Hu et al. 2010; Lin et al. 2010; Dolique et al. 2010; Zhang et al. 2010; Singh et al. 2011; Chen et al, 2005; Chuang et al. 2011; Lin and Tsai 2011; Liu et al. 2012; Manzoni et al. 2013; Li et al. 2013; Qiu 2013; Tariq et al. 2013; Daoud et al. 2013; Pradeep et al. 2013; Manzoni et al. 2013; Hsu et al. 2005; Chen et al. 2006; Yeh et al. 2007; Hsu et al. 2007; Wang et al. 2007; Tung et al. 2007; Wang et al. 2008), published in 2000–2015 years the results of the studying the methods for manufacturing high-entropy alloys (HEA) of different chemical composition, microstructure and properties are considered. It is necessary to add the papers (Tsai et al. 2010; Wen et al. 2009; Strife and Passoja 1980; Ng et al. 2012; Jones et al. 2014; Shun and Du 2009; Kao et al. 2009; Tsai et al. 2009; Otto et al. 2013; Gludovatz et al. 2014; Mills 1997) in which the effect of thermal and deformation treatment on the structure and mechanical properties of HEA alloys were analysed. The original results obtained in the field of HEA prior to 2015 are considered in detail in analytical reviews (Zhang et al. 2014; Cantor 2014; Miracle and Senkov 2017; Zhang and Zhang 2018; Osintsev et al. 2021) where the HEA thermodynamics is described, results of modelling of their structure are considered and new variants of methods for obtaining the multi-component alloys are discussed.

The HEA studies have shown that it is possible to form in them nanodimensional structures and even amorphous phases due to considerable distortions of lattice caused by the difference in the atomic radii of substitution elements. In this case, the rate of diffusion processes decreases, and as a result, the speed of crystal growth reduces (Pogrebnyak et al. 2014).

Due to differences in dimensions of atoms of different metals the HEA crystal lattice turns out to be heavily distorted, therefore the structure of such phases may be considered as intermediate between stable crystal phases with relatively small equilibrium concentration of defects, including impurity atoms, and metastable metal glasses in which long-range order is completely absent.

By now more than 10,000 papers in Scopus and Web of Science bases had already been published. The share of publications on HEA amounts from 5% in Iran to 20–22% in China and the USA (Rogachev 2020). Such an exponential growth of publications will not fail to raise the question: whether the HEA concept is just another scientific fashion such as the ideas of individual dislocations of the last century capable of explaining all the diversity of deformation behaviour of crystalline materials.

By now there is no unambiguous answer to this question. It is connected with the fact that direct comparison of data is difficult, due to differences in the type and concentration of principal elements, the type and extent of thermo-mechanical processing, and the temperature and duration of post-process thermal treatment (George et al. 2020).

Practically all types of such alloys (structural, cryo- and heat resistant, corrosion-resistant, those with special magnetic and electrical properties), as well as compounds (carbides, nitrides, oxides, borides, silicides), are being developed. In the majority of cases, researchers succeed in manufacturing a single-phase high-entropy material or multi-phase material consisting of a multi-component matrix and inclusions which may result in dispersion hardening (Rogachev 2020).

Nowadays the process of accumulating and comprehending the results concerning HEA production, their mechanical properties, microstructure, etc. goes on beginning with classical alloys of Cantor CrMnFeCoNi and Senkov TiZrHfNbTa and finishing by unique compositions with rare-earth elements. Estimation of HEA uniqueness in comparison with traditional alloys is a crucial importance for the development of different branches of industry. On the basis of the data available, there are good reasons to consider that HEA's rapid development will continue in the nearest future. The major task of the present monograph is to analyse the latest results over the last years according to the main sections: 1. Methods of HEA production; 2. Mechanical properties and mechanisms of deformation; 3. Stability; 4. Prospects for application; 5. Possibilities of use of external energy effects for improving the HEA structural-phase states and properties.

Precisely these sections are decisive, in our opinion, in estimating the perspectives of the large-scale industrial introduction of high-entropy alloys into the industry.

References

Braic M., Braic V., Balaceanu M., Zoita C.N., Vladescu A., Grigore E.: Surf. Coatings Technol. **204**, 2010 (2010)
Cantor B.: Entropy **16**, 4749 (2014)
Chen M.-R., Lin S.-J., Yeh J.-W., Chen S.-K., Huang Y.-S., Tu C.-P.: Mater. Trans. **47**, 1395 (2006)
Chen M.-R., Lin S.-J., Yeh J.-W., Chuang M.-H., Lee P.-H., Huang Y.-S.: Metall. Mater. Trans. A-Physical Metall. Mater. Sci. - Met. MATER TRANS A **37**, 1363 (2006)
Chen Y.Y., Duval T., Hung U.D., Yeh J.W., Shih H.C.: Corros. Sci. **47**, 2257 (2005)
Chuang M.-H., Tsai M.-H., Wang W.-R., Lin S.-J., Yeh J.-W.: Acta Mater. **59**, 6308 (2011)
Daoud H.M., Manzoni A., Völkl R., Wanderka N., Glatzel U.: JOM **65**, 1805 (2013)
Dolique V., Thomann A.-L., Brault P., Tessier Y., Gillon P.: Surf. Coatings Technol. **204**, 1989 (2010)
George E.P., Curtin W.A., Tasan C.C.: Acta Mater. **188**, 435 (2020)
Gludovatz B., Hohenwarter A., Catoor D., Chang E.H.E.H., George E.P.E.P., Ritchie R.O.R.O.: Science (80-.). **345**, 1153 (2014)
Hsu U.S., Hung U.D., Yeh J.W., Chen S.K., Huang Y.S., Yang C.C.: Mater. Sci. Eng. A **460–461**, 403 (2007)
Hsu Y.-J., Chiang W.-C., Wu J.-K.: Mater. Chem. Phys. **92**, 112 (2005)
Hu Z., Zhan Y., Zhang G., She J., Li C.: Mater. Des. **31**, 1599 (2010)
Huang P.-K., Yeh J.-W.: Surf. Coatings Technol. **203**, 1891 (2009)
Jones N.G., Frezza A., Stone H.J.: Mater. Sci. Eng. A **615**, 214 (2014)
Kao Y.-F., Chen T.-J., Chen S.-K., Yeh J.-W.: J. Alloys Compd. **488**, 57 (2009)
Li A., Zhang X.: Acta Metall. Sin. (English Lett. **22**, 219 (2009)
Li B., Peng K., Hu A., Zhou L., Zhu J., Li D.: Trans. Nonferrous Met. Soc. China **23**, 735 (2013)
Lin C.-M. Tsai H.-L.: Intermetallics **19**, 288 (2011)
Lin M.-I., Tsai M.-H., Shen W.-J., Yeh J.-W.: Thin Solid Films **518**, 2732 (2010)
Liu L., Zhu J B., Zhang C., Li J.C., Jiang Q.: Mater. Sci. Eng. A **548**, 64 (2012)
Manzoni A., Daoud H., Mondal S., van Smaalen S., Völkl R., Glatzel U., Wanderka N.: J. Alloys Compd. **552**, 430 (2013)
Manzoni A., Daoud H., Völkl R., Glatzel U., Wanderka N.: Ultramicroscopy **132**, 212 (2013)
Mills W.J.: Int. Mater. Rev. **42**, 45 (1997)
Miracle D.B., Senkov O.N.: Acta Mater. **122**, 448 (2017)
Ng C., Guo S., Luan J., Shi S., Liu C.: Intermetallics **31**, 165 (2012)
Osintsev K.A., Gromov V.E., Konovalov S.V., Ivanov Y.F., Panchenko I.A.: Izv. Ferr. Metall. **1** (2021)
Otto F., Dlouhý A., Somsen C., Bei H., Eggeler G., George E.P.: Acta Mater. **61**, 5743 (2013)
Pogrebnyak A.D., Bagdasapyan A.A., Yakushchenko I.V., Beresnev V.M.: Russ. Chem. Rev. **83**, 1027 (2014)
Pradeep K.G., Wanderka N., Choi P., Banhart J., Murty B.S., Raabe D.: Acta Mater. **61**, 4696 (2013)
Rogachev A.S.: Phys. Met. Mater. Sci. **121**, 807 (2020)
Shun T.-T., Du Y.-C.: J. Alloys Compd. **478**, 269 (2009)
Singh S., Wanderka N., Murty B.S., Glatzel U., Banhart J.: Acta Mater. **59**, 182 (2011)
Strife J., Passoja D.: Metall. Trans. **11**, 1341 (1980)
Tariq N.H., Naeem M., Hasan B.A., Akhter J.I., Siddique M.: J. Alloys Compd. **556**, 79 (2013)
Tong C.-J., Chen M.-R., Yeh J.-W., Lin S.-J., Lee P.-H., Shun T.-T., Chang S.-Y.: Metall. Mater. Trans. A-Physical Metall. Mater. Sci. - Met. MATER TRANS A **36**, 1263 (2005a)
Tong C.-J., Chen Y.-L., Yeh J.-W., Lin S.-J., Lee P.-H., Shun T.-T., Tsau C.-H., Chang S.-Y.: Metall. Mater. Trans. A **36**, 881 (2005b)
Tsai C.-W., Chen Y.-L., Tsai M.-H., Yeh J.-W., Shun T.-T., Chen S.-K.: J. Alloys Compd. **486**, 427 (2009)
Tsai C.-W., Tsai M.-H., Yeh J.-W., Yang C.-C.: J. Alloys Compd. **490**, 160 (2010)
Tung C.-C., Yeh J.-W., Shun T., Chen S.-K., Huang Y.-S., Chen H.-C.: Mater. Lett. **61**, 1 (2007)
Wang X.F., Zhang Y., Qiao Y., Chen G.L.: Intermetallics 15, **357** (2007)

Wang Y.P., Li B.S., Ren M.X., Yang C., Fu H.Z.: Mater. Sci. Eng. A **491**, 154 (2008)
Wen L.H., Kou H.C., Li J.S., Chang H., Xue X.Y., Zhou L.: Intermetallics **17**, 266 (2009)
X.-W. Qiu.: J. Alloys Compd. 555, **246** (2013)
Yeh J.W., Chen Y.L., Lin S.J., Chen S.K.: Mater. Sci. Forum **560**, 1 (2007)
Zhang K. B., Fu Z.Y., Zhang J.Y., Shi J., Wang W.M., Wang H., Wang Y.C., Zhang Q.J.: J. Alloys Compd. **502**, 295 (2010)
Zhang W., Zhang Y.: Sci. China Earth Sci. **2** (2018)
Zhang Y., Zuo T.T.T.T., Tang Z., Gao M.C., Dahmen K.A.K.A., Liaw P.K.P.K., Lu Z.P.Z.P.: Prog. Mater. Sci. **61**, 1 (2014)

Contents

Chapter 1
Methods of Manufacturing the High-Entropy Alloys

Abstract In this chapter, the most promising techniques of the manufacturing of high entropy alloys are considered. Generally, methods could be divided by manufacturing of powders and corresponding powder metallurgy, casting techniques, coatings deposition and additive manufacturing methods of the high-entropy alloys. All of the mentioned methods are described in detail, both disadvantages and advantages are given. Technologies are ranged by the quantity of the published articles over the whole period up to December 2020. To determine the most popular manufacturing methods and to process the high-entropy alloys, the analysis of a number of published works was carried out, including papers, books and materials of conferences over the whole period up to December 2020 inclusively. Statistical data were collected from Scopus and Web of Science (WoS) databases, combined the most similar methods and presented as histograms. Based on this analysis, it could be drawn that one of the most widespread methods of obtaining bulk samples nowadays is vacuum arc melting technology. Manufacturing the coatings is performed using laser deposition in the majority of the presented works. The most widespread method of additive manufacturing is selective laser melting. The Research is supported by the Russian Science Foundation (project No. 20-19-00452).

The high-entropy alloys, as a new class of materials, appeared at the beginning of twenty-first century. In the first investigations concerning the study of microstructure and properties of high-entropy alloys (HEA), the induction and arc melting followed by casting (Cantor et al. 2004; Yeh et al. 2004) were used as methods of production. Later on, the number of methods increased, however, common to all remained that the majority of them use powders as initial materials.

In the chapter the different technologies of manufacturing the high-entropy alloys are considered, their characteristic features are presented as well as their advantages and disadvantages are revealed. In order to determine the most popular methods of manufacturing and processing the high-entropy alloys the analysis of the number of published works was carried out including papers, books and materials of conferences over the whole period up to December 2020, inclusively. Statistical data were collected from Scopus and Web of Science (WoS) databases, combined with the most similar methods and presented as histograms.

V. E. Gromov et al., *Structure and Properties of High-Entropy Alloys*, Advanced Structured Materials 107, https://doi.org/10.1007/978-3-030-78364-8_1

1.1 Manufacturing of HEA Powders

Consider the technologies whereby the powders of high-entropy alloys are manufactured. Figure 1.1 presents statistics of published works wherein the different methods of obtaining the high-entropy alloys powders were used. It demonstrates that the most popular, for the present moment, is the technology of mechanical alloying/high-energy ball milling. A number of published works using the method amount to 621 in the Scopus database and 3604 in Web of Science. The second and the third places in popularity belong to the technology of gas atomisation and water atomisation.

Mechanical Alloying of High-Entropy Alloys' Powder

Mechanical alloying is a technology of powder treatment that enables the chemically and structurally homogeneous materials to be obtained, beginning with mixtures of elemental powders (with high purity > 99.5%). In mechanical alloying, the powder particles are gripped between milling bodies and subjected to the process of deformation and/or failure, and final chemical homogeneity depends on the mechanical behaviour of initial powder components. The process can be done in planetary mills wherein there are 3 or 4 drums revolving about the central axis and simultaneously about their own axes in opposite direction (similar to the motion of planets about the Sun) (Fig. 1.2a). The particles of the material being ground undergo a large number of collisions with milling bodies (balls) and drum walls (Fig. 1.2b) (Chernik et al. 2007). Such substances as ethanol, toluene, methanol, benzene, heptane, etc. (Shivam et al. 2018; Ganesh and Raghavendra 2020) can be used in order to avoid joining and cold welding of powder particles. The controlled parameters in the method are the following: mixing time, speed of rotation of the planetary disk, type and dimension of milling bodies, process control agent, as well as shielding gas.

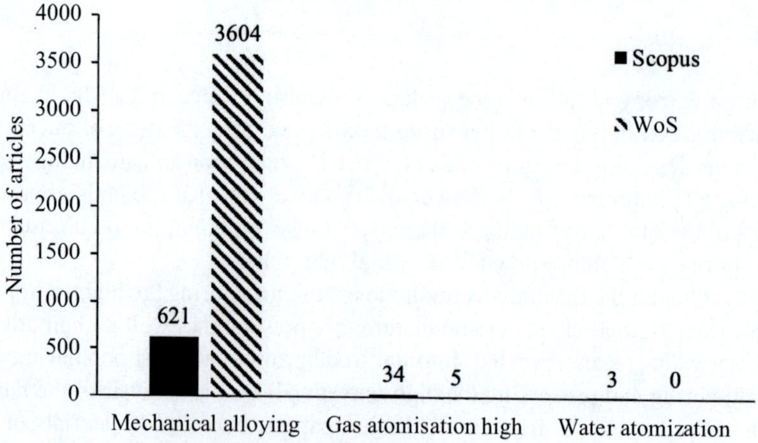

Fig. 1.1 Distribution histogram of the number of works published up to December 2020, depending on the method of manufacturing HEAs powders

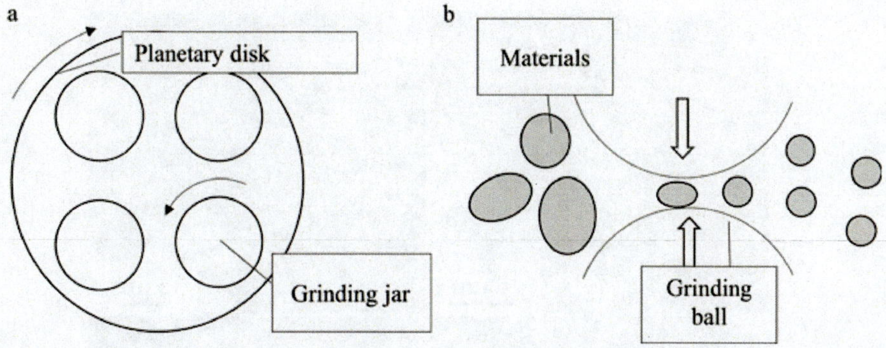

Fig. 1.2 Diagram of planetary mill arrangement (**a**); **b**—the process of mechanical alloying of powders (the arrows show the direction of motion of milling bodies)

In the research (Shivam et al. 2018) the time effect of mechanical allowing on particles' size and deformation of the crystal lattice of AlCoCrFeNi equiatomic high-entropy alloy powder was studied. It was found that in mechanical alloying for 5 h the crystallites' size amounted to 19 nm, and deformation was 0.56%. With every increase in mechanical alloying time by 5 h the decrease in grain size by 2–3 nm takes place and approaches the least value for 30 h (Fig. 1.3). In this case, deformation increases linearly and reaches a maximum of 1.2% for 30 h.

Advantages: the method enables a high homogeneity of alloy composition to be achieved.

Disadvantages: the necessity for using the process control agents, contaminating a powder composition. Possible material losses related to friction.

Gas atomisation of High-Entropy Alloy Powders

Gas atomisation technology consists of melting powders of pure substances and subsequent passing a pressurized gas coolant through a liquid metal (Fig. 1.4) (Alshataif et al. 2019). Argon is usually used as a gas. In order to ensure a softer process of atomisation, a nozzle preheating to 700 °C can be used as it was performed in the research (Cheng et al. 2019) in which a powder of AlCoCrFeNi equiatomic high-entropy alloy was manufactured. The dimensions of obtained particles varied in the range of 10–250 μm (Fig. 1.5).

Advantages: it enables the powder particles of spherical shape with a smooth surface to be obtained. It, in its turn, ensures a better density and looseness to powders.

Disadvantages: particles have different sizes which can differ by 200 μm.

Production of High-Entropy Alloys Powders by Means of Water Atomisation

Particles' shape of powders produced by the water atomisation method differs considerably from those obtained by gas atomisation (Fig. 1.6). Changes in powder morphology may be connected with the fact that the heat conductivity of gas coolant is less in comparison with that of a liquid. Because of this the powders produced by spraying in gas jet cool more slowly that gives more time to decrease the sizes of

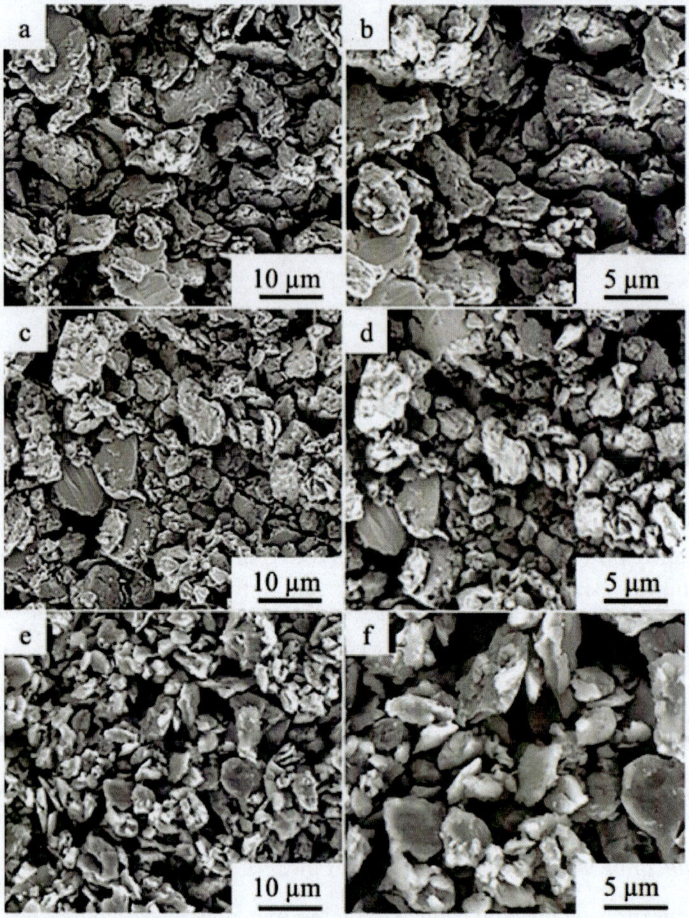

Fig. 1.3 SEM of AlCoCrFeNi powders manufactured by mechanical alloying for **a**, **b** 10 h, **c**, **d** 20 h, **e**, **f** 30 h (Shivam et al. 2018)

Fig. 1.4 Schematic of the working principle of the powder atomisation process (Alshataif et al. 2019)

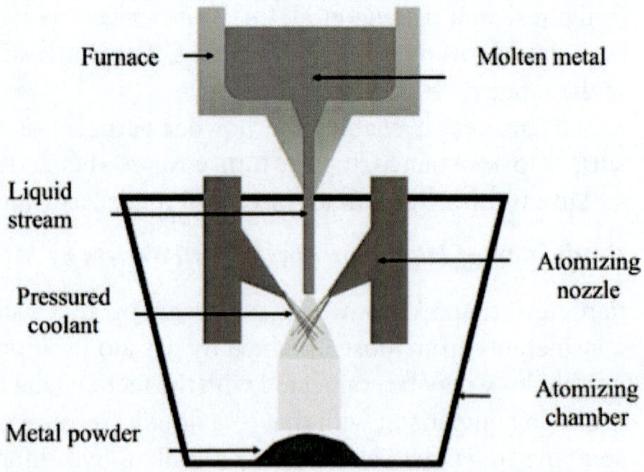

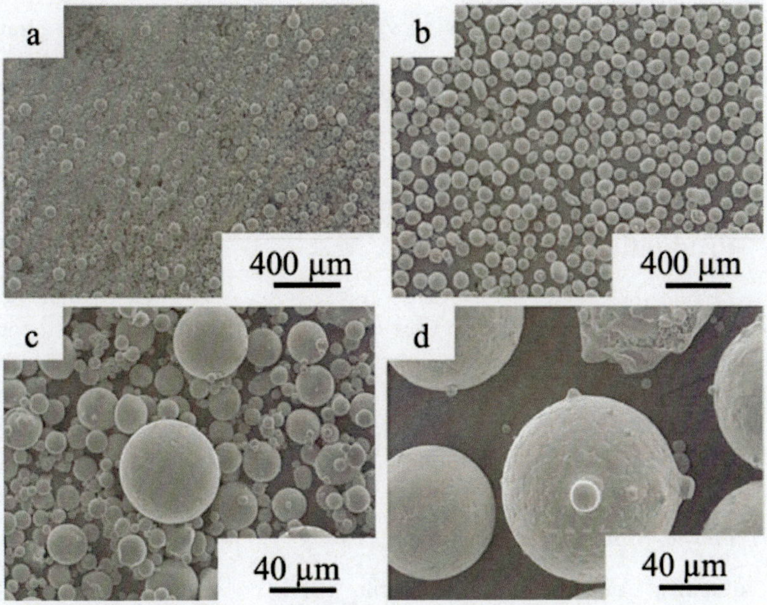

Fig. 1.5 SEM-image showing the morphology of obtained powders AlCoCrFeNi with different particle dispersion: **a**, **b** 10–60 μm and **c**, **d** 60–90 μm (Cheng et al. 2019)

Fig. 1.6 SEM of CoCrFeMnNi powder produced by water atomisation method (Yim et al. 2019)

particles and lends a spherical shape to them, but powder particles obtained in the liquid medium have an irregular shape and large sizes. So, in water atomisation under pressure 30 bar the powders with the size of agglomerated particles in the range of 150–200 μm (Fig. 1.6) (Yim et al. 2019) were obtained from the melt consisting of Co, Cr, Fe, Mn and Ni.

Advantages: sizes of particles vary in smaller range than in gas atomisation (50 μm).

Disadvantages: particles are of irregular shape.

1.2 Methods of Powder Metallurgy Used for HEAs Production

As a result of analysis of the powder metallurgy methods being used for HEAs production, it was determined that the most frequently used methods are those belonging to hot pressing: spark plasma smelting, hot pressing by indirect resistance heating and hot isostatic pressing (Fig. 1.7). Liquid-phase sintering, self-propagating high-temperature synthesis (SHS) and cold pressing are less used for HEA production.

Spark Plasma Sintering

The equipment for spark plasma sintering (SPS) consists of the following main parts: agglomeration press with a vertical single-axis mechanism of pressurisation, specially developed punch electrodes including a water cooler, a vacuum chamber with water cooling, a mechanism of the atmosphere (vacuum/air/argon gas) control, a special DC pulsed sintering power generator, a control unit of cooling water, a measuring and controlling unit of Z-axis position, temperature measuring and

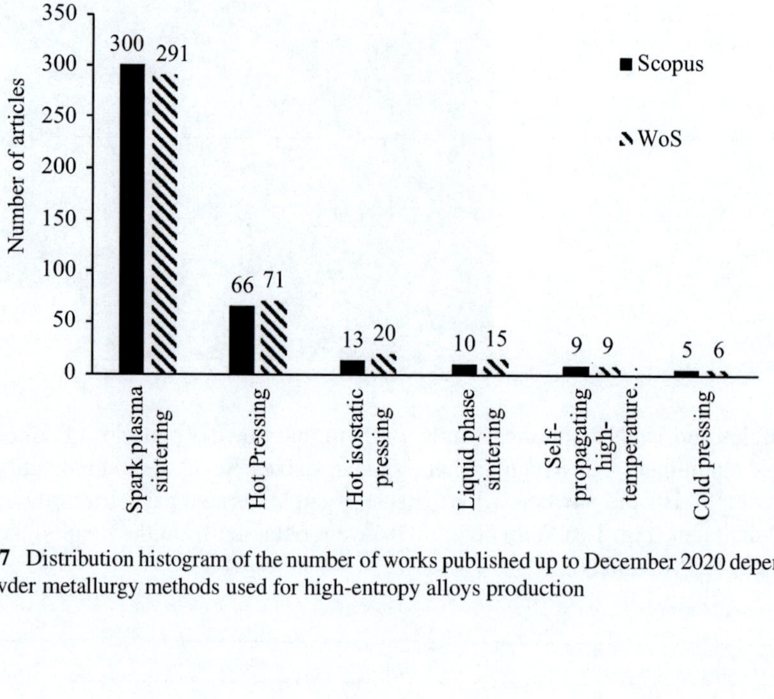

Fig. 1.7 Distribution histogram of the number of works published up to December 2020 depending on powder metallurgy methods used for high-entropy alloys production

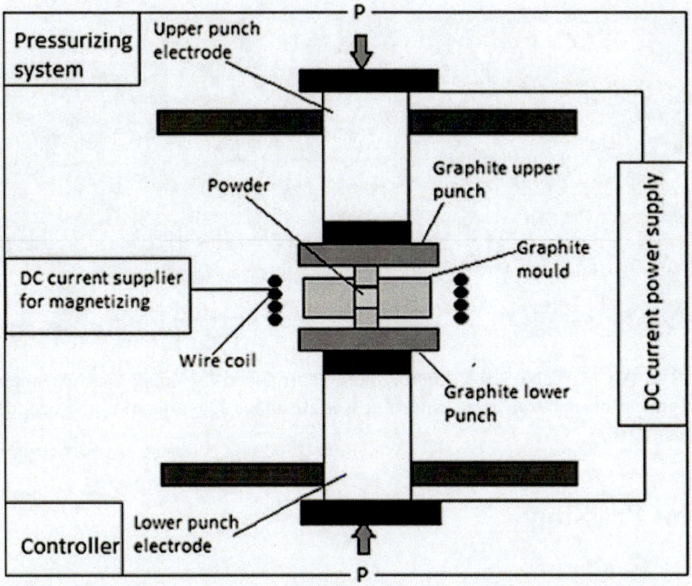

Fig. 1.8 Block diagram of spark plasma sintering process (Mohammed et al. 2021)

controlling unit, unit of applied pressure display and different safety interlock devices (Fig. 1.8) (Mohammed et al. 2021).

The process of spark plasma sintering is a sintering method during which external pressure and electric field are simultaneously applied to intensify a powder mixture densification. In this process, a can current is passed through a conducting matrix and a sample, and it creates an electric field in the sintering process. Sintering occurs at the expense of powder heating both from the outside and from the inside (Cavaliere and in Spark 2019).

In the research (Zhang et al. 2016) the AlCoCrFeNi equiatomic high-entropy alloy was manufactured by means of spark plasma sintering. Powder mixture prepared preliminary in mixer for 5 h was charged to a graphite matrix and heated from room temperature to 1200 °C at a heating rate of 50 °C/min. Then sintering was done at the temperature for 20 min under the pressure of 30 MPa followed by cooling in the furnace. It was found that the resultant material consisted of FCC-phase and duplex BCC-structure consisting of disordered and ordered BCC-phases (Fig. 1.9). The obtained HEA demonstrated excellent mechanical properties as a result of the BCC- and FCC-phase combination.

Advantages: heating is carried out both from the outside and from the inside of powders that increases economic effectiveness because it decreases a sintering temperature and reduces the time for its maintaining.

Disadvantages: powders should conduct electric current.

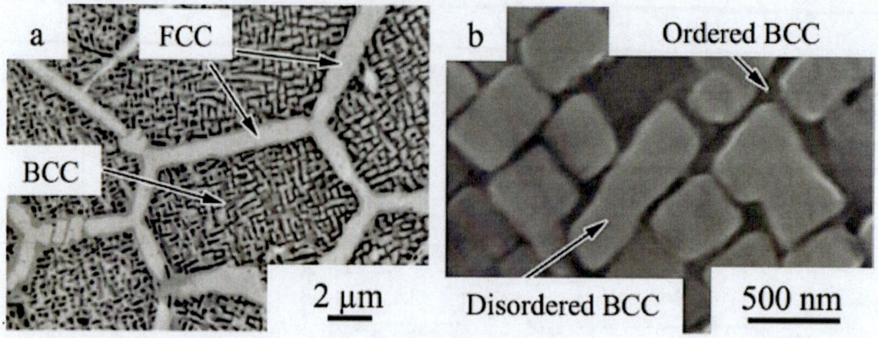

Fig. 1.9 The AlCoCrFeNi HEA microstructure produced by spark plasma sintering: **a** low-magnified image of electron backscattering, **b** high image magnification in secondary electrons (Zhang et al. 2016)

1.3 Hot Pressing

The process of hot pressing can be carried out in a vacuum or controlled gas atmosphere. The prepared powder is charged in a chamber that is heated with graphite heating elements and obtains heat from it by convection (Fig. 1.10). The method is simple in comparison with the methods of spark plasma sintering and hot isostatic pressing, it enables a high heating temperature to be provided and is independent of powder conductivity. Owing to the homogeneity of temperature distribution in a mould the large samples in diameter up to 150 mm (Hu et al. 2014) may be produced.

Fig. 1.10 Diagram of a hot pressing process (Hu et al. 2014)

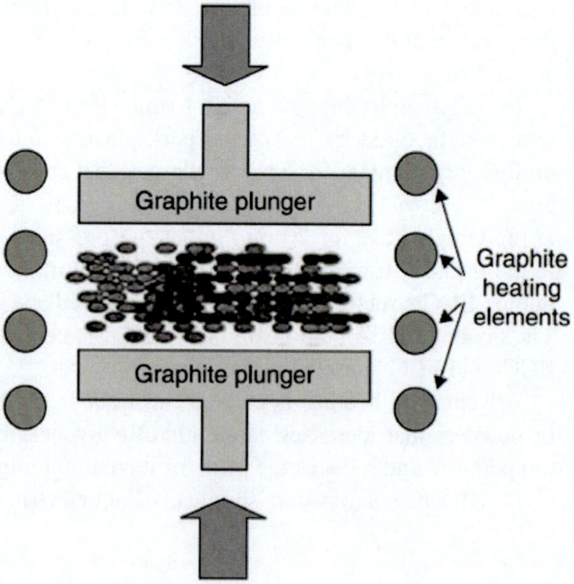

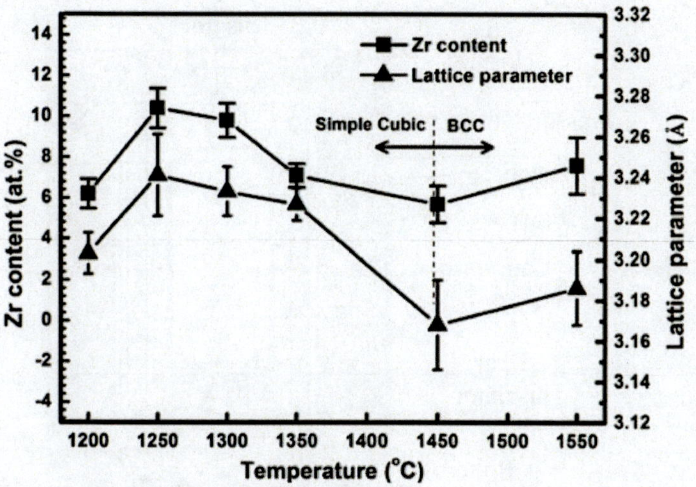

Fig. 1.11 Effect of hot pressing temperature on lattice parameter and Zr content in matrix phase (Tan et al. 2016)

As it was shown in the research (Tan et al. 2016) the increase in hot pressing temperature affects the phase composition by changing, in this case, the lattice parameter and chemical composition of $Al_2NbTi_3V_2Zr$ high-entropy alloy. With the increase in sintering temperature, the phase composition of the matrix varies from simple cubic lattice to BCC-crystal lattice. In this case, a correlation between Zr content and lattice parameter of matrix phase was detected because Zr had the largest radius among five atoms. The effect of sintering temperature on Zr content and lattice parameter of matrix phase is presented in Fig. 1.11. The Zr content and lattice parameter increased initially and then decreased at sintering temperatures of 1200–1350 °C (Tan et al. 2016).

Advantages: a simple method that enables the powders to be sintered independent of their electrical conductivity.

Disadvantages: low energy and economical effectiveness because the method requires much time to heat a chamber.

Hot Isostatic Pressing

A distinguishing feature of hot isostatic pressing (HIP) is that the sintering process takes place under constant pressure. In contrast to hot pressing the pressure affects uniformly in all directions (Fig. 1.12). The main advantage of the method is a large variety of sample shapes that may be subjected to compacting. Sintering process compacts and binds powders or porous bodies simultaneously. The process can improve mechanical properties and material workability (Hu et al. 2014).

In the research (Joseph et al. 2018) hot isostatic pressing was applied to samples of $Al_xCoCrFeNi$ high-entropy alloy produced with direct laser deposition. It was established that $Al_{0.6}CoCrFeNi$ alloy treated by hot isostatic pressing showed a considerably varied structure with enlarged globular grains of FCC- and BCC-phases

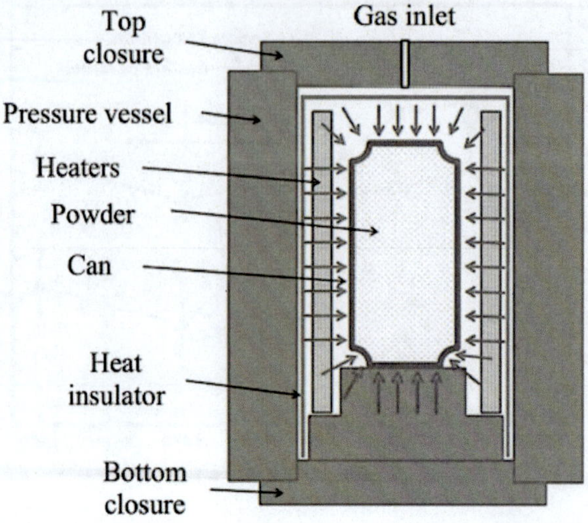

Fig. 1.12 Diagram of a hot isostatic pressing process (Torralba et al. 2014)

of different dimensions and morphology. Initial and treated samples demonstrated similar mechanical behaviour under compression, however, under tension the treated sample showed a brittle failure. It is supposed that the decrease in plasticity under the tension of $Al_{0.6}$ alloy after hot isostatic pressing is related to the formation of coarse inclusions of B_2-phase at interphase and grain boundaries (Fig. 1.13).

Advantages: pressing occurs in all directions simultaneously that makes it possible to use a large number of different sample shapes.

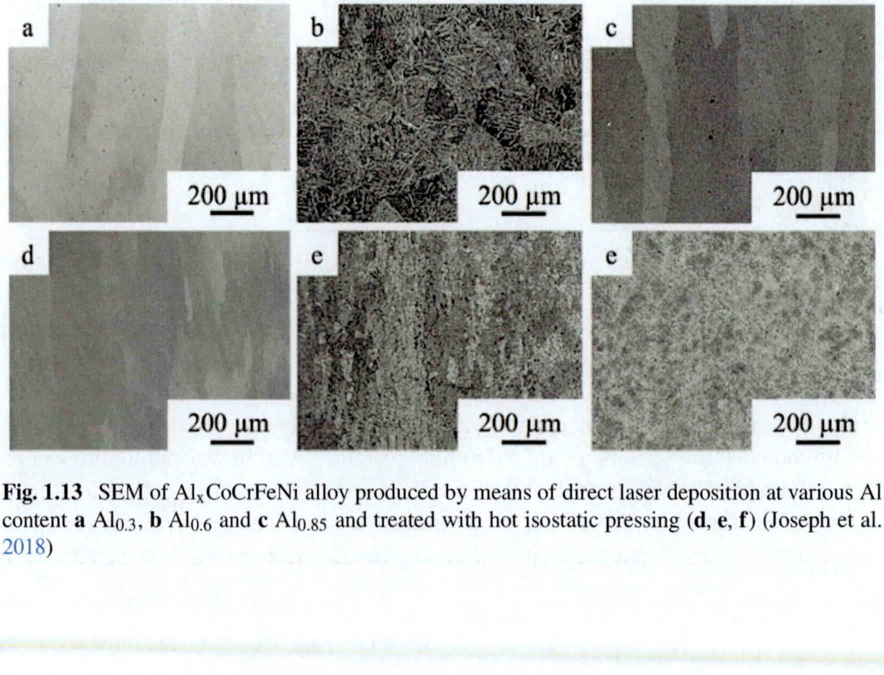

Fig. 1.13 SEM of $Al_xCoCrFeNi$ alloy produced by means of direct laser deposition at various Al content **a** $Al_{0.3}$, **b** $Al_{0.6}$ and **c** $Al_{0.85}$ and treated with hot isostatic pressing (**d**, **e**, **f**) (Joseph et al. 2018)

Disadvantages: the technology is limited by its complexity and high equipment cost. The highest working pressure amounts to 200 MPa.

Liquid-Phase Sintering

A process of liquid-phase sintering begins with mixing the powder particles having different melting temperature range. On heating, the powder particles of one type remain in a solid-state while the particles of other powder form a liquid phase. It results in the rearrangement of particles remaining in the solid phase, the change in their shape and subsequent densification. Diffusion of solid particles in liquid results in the elimination of pores (Fig. 1.14).

In the research (Jian et al. 2020) the method was employed to form a composite material whose filling compound was tungsten particles and matrix was $Al_{0.5}Cr_{0.9}FeNi_{2.5}V_{0.2}$ high-entropy alloy. The process of liquid-phase sintering was performed in H_2 atmosphere at a temperature of 1500 °C for 0, 30, 60, 90 and 120 min, respectively (Fig. 1.15).

A composite material obtained as a result of liquid-phase sintering is of fine-grained structure with the tungsten grain size of 9.7 μm and density 16.1 g/cm^3 supposing a dissolution of 4% tungsten. The results open a new possibility for producing fine-grained tungsten alloys due to HEA application as a matrix.

Advantages: the method enables the composite materials of complex composition to be produced.

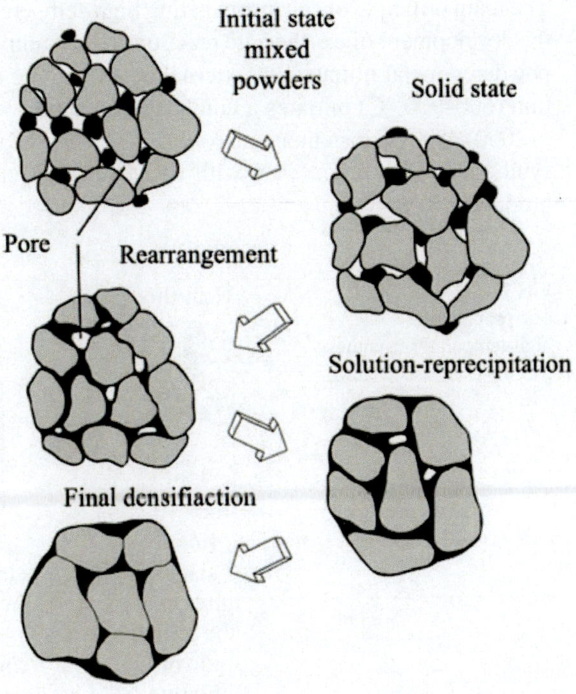

Fig. 1.14 Diagram of microstructural change in liquid-phase sintering (German et al. 2009)

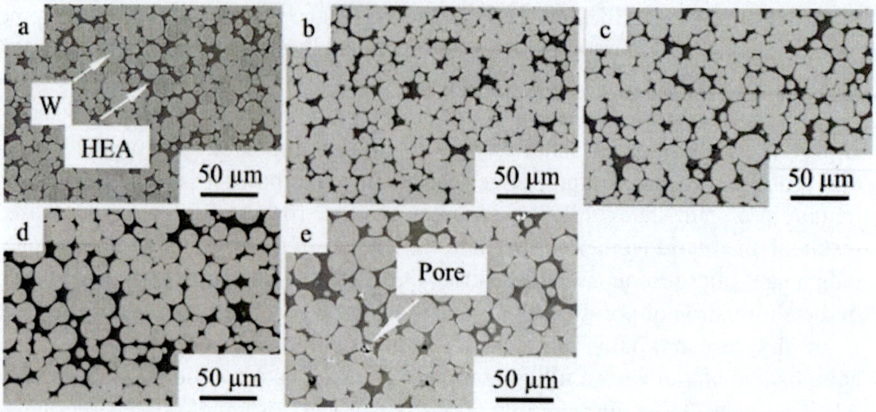

Fig. 1.15 Microstructure of W-$Al_{0.5}Cr_{0.9}FeNi_{2.5}V_{0.2}$ alloy depending on holding time **a** 0 min, **b** 30 min, **c** 60 min, **d** 90 min **e** 120 min, respectively. **f** Typical SAED sample of HEA matrix W-HEA for holding time 0 min (Jian et al. 2020)

Disadvantages: high distortion of the crystal lattice, deterioration of mechanical properties due to solidification of brittle phases along grain boundaries and/or grain growth in sintering, limitation of finished part for high-temperature applications.

Self-propagating High-Temperature Synthesis

The main principle of self-propagating high-temperature synthesis (SHS) consists in the development of exothermic reactions (for example, redox, oxidation, thermal) in powdered metal initiated by external heat flux (Fig. 1.16). The temperature of ignition (600–900 °C) initiates a combustion reaction which increases the temperature to 2000–3500 °C instantaneously during several seconds ensuring a self-propagating synthesis of the entire part with the result that sintered material (Nee 2015) is produced.

Fig. 1.16 Diagram of self-propagating high-temperature synthesis process (Nee 2015)

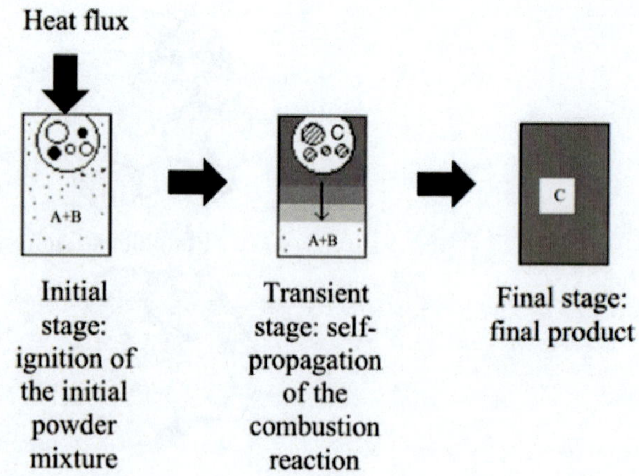

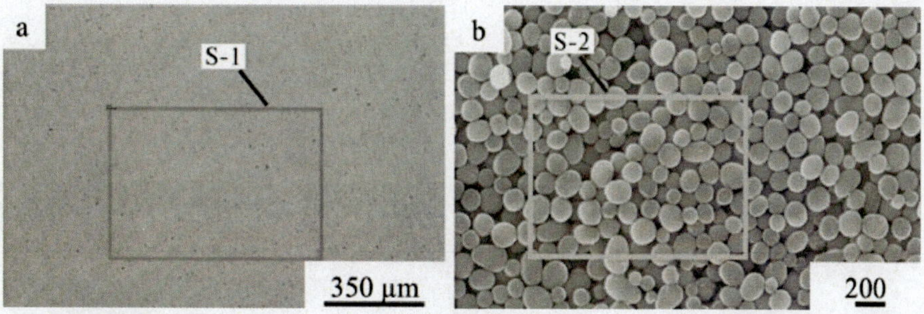

Fig. 1.17 Microstructure of $NiCrCoFeMnAl_{1.6}$ alloy produced by means of self-propagating high-temperature synthesis **a** prior to and **b** after etching (Sanin et al. 2016)

As it was shown in the research (Sanin et al. 2016) self-propagating high-temperature synthesis permitted a formation of the high-entropy alloy of Ni–Cr–Co–Fe–Mn–Al_x system with composite structure and nanodimensional disperse inclusions (Fig. 1.17).

Data on integral chemical composition allow the conclusion that the composition structure of $NiCrCoFeMnAl_{1.6}$ alloy is formed by NiAl, but numerous rounded nanodimensional inclusions with sizes of 100–150 nm consist of solid solution based on Cr–Fe–Co–Mn polymetallic melt.

Advantages: a simple process, low energy consumption, high quality of products (i.e. chemical purity), simultaneous formation and densification of material as well as a large range of attainable microstructures (Nee 2015).

Disadvantages: the SHS reactions are very difficult to control because initiation and termination of reaction proceed during a very short time interval. Reaction kinetics depends on several parameters of the process and a little deviation of parameters may change in kinetics resulting in a presence of unreacted phases in a product that is undesirable. Products manufactured by the SHS method can generate metastable phases that may also be undesirable (Mishra and Pathak 2009).

Cold Pressing

Cold pressing is a process that consists of pressing powdered particles at room temperature (Fig. 1.18). In the research (Fan et al. 2013) the samples of high-entropy alloy were produced and a relative density of more than 91% was achieved by means of cold pressing at a load of 310 MPa.

The advantages of the method consist in its simplicity and economical efficiency. The disadvantage of the method is that it produces non-uniform densification.

1.4 Technologies of Casting Used for HEA Production

Another group of technologies by means of which high-entropy alloys may be produced is casting technologies. The most frequently used is vacuum arc melting (Fig. 1.19). Then go vacuum induction melting and directional solidification which

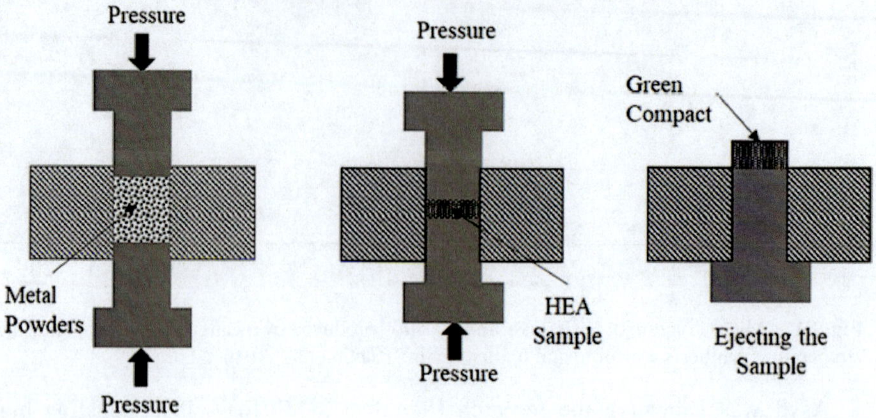

Fig. 1.18 Diagram of the cold pressing process (Alshataif et al. 2019)

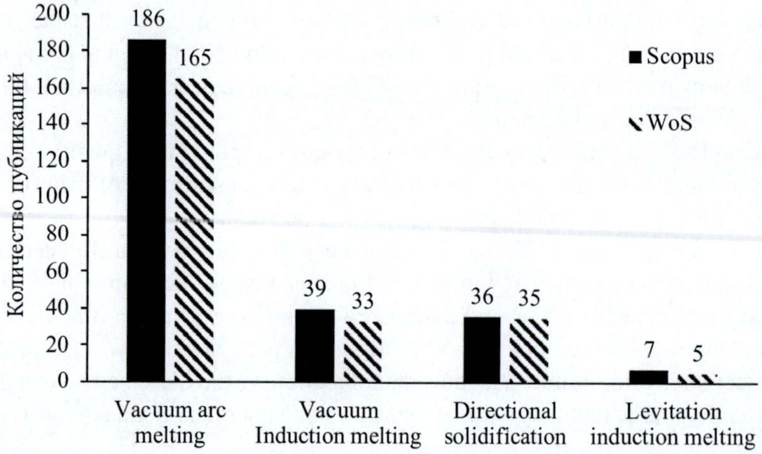

Fig. 1.19 Distribution histogram of the number of works published up to December 2020 depending on casting technologies used for high-entropy alloys' production

is mentioned in present-day published works equally often as methods of HEA production.

Vacuum Arc Melting

In vacuum-arc melting an electric arc furnace and mechanical pump, which enables one to pump out air fully and reach a pressure of $6 \bullet 10^{-2}$ Pa, are used. In the production of high-entropy alloys with the method, the initial powders are usually remelted no less than five times in order to reach homogeneity of composition (Shun and Hung 2018). By means of vacuum arc melting the samples of $AlCoCr_xFeNi$ alloy with different Cr content were produced. It is found that with the increase in Cr content

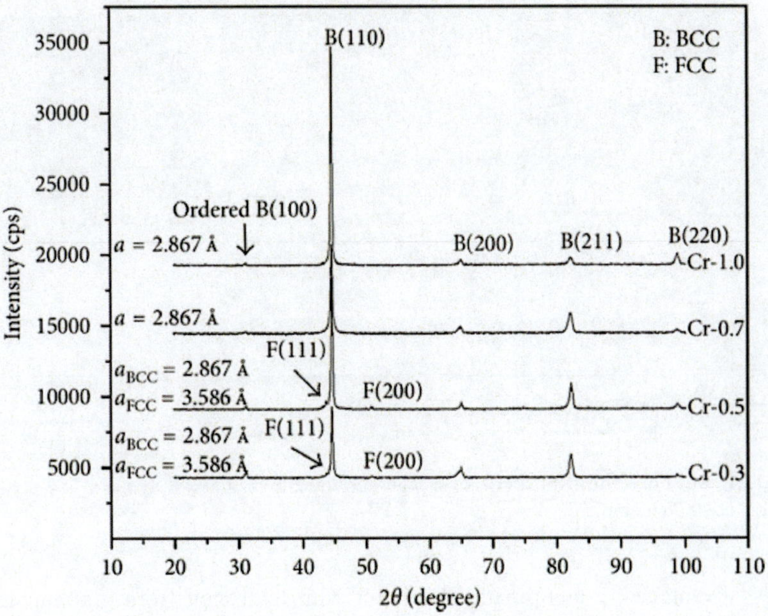

Fig. 1.20 Diffraction patterns of $AlCoCr_xFeNi$ alloy with different Cr content (Shun and Hung 2018)

the BCC-phase fraction increases, and the structure transfers from mixed BCC + FCC to a single-phase BCC crystal lattice (Fig. 1.20).

Advantages: vacuum arc melting method is simple and allows the samples free from impurities of different gases to be produced.

Disadvantages: evaporation of low-melting elements in the process of melting, formation of heterogeneous structure which forms due to low rate of crystallisation.

Vacuum Induction Melting

In vacuum induction melting an electromagnetic field producing eddy currents in powdered metal is used as the energy source that results in its melting. In the research (Wu et al. 2020) induction melting was performed at a vacuum level of $5 \cdot 10^{-3}$ Pa to produce a matrix composite from high-entropy alloy. The induced current was initially set to 600 A and decreased to 300 A prior to mixing. Then the molten metal was poured into the copper water-cooled crucible and cooled to room temperature. It was established that matrix composites of NbC/FeCrNiCu high-entropy alloys were composed of NbC fine particles embedded into FCC-matrix. Copper segregates on grain boundaries. The NbC particles have rectangular morphology with an average size of 1.6 μm in composites $(NbC)_5$ (Fig. 1.21).

Advantages: the method of vacuum induction melting is simple and allows the samples free from impurities of different gases to be produced. Vacuum being formed on melting protects the material from oxidation.

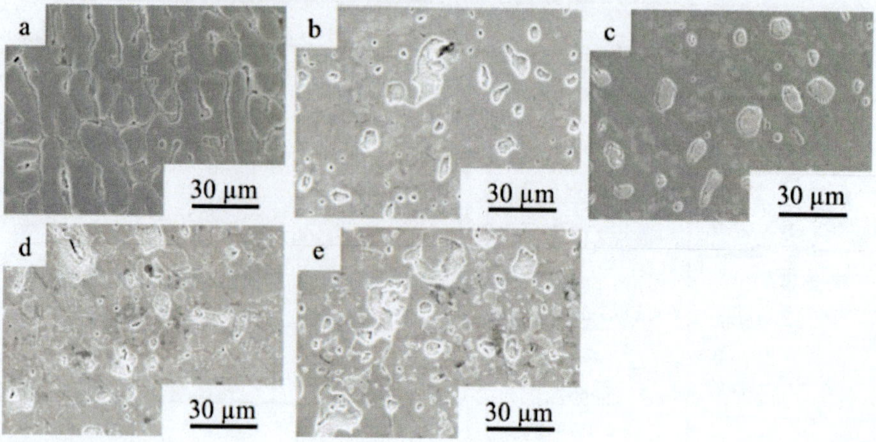

Fig. 1.21 SEM of xvol%NbC/FeCrNiCu high-entropy alloys: **a** $x = 0$; **b** $x = 2.5$; **c** $x = 5$; **d** $x = 7.5$; **e** $x = 10$ (Wu et al. 2020)

Disadvantages: contamination of ingot with hydrogen from graphite crucible. Loss of elements as a result of evaporation is possible.

Directional Solidification

Manufacture of castings by methods of directional solidification is based on the movement of ceramic mould filled with molten metal at a definite rate from the heating zone to the cooling area. Thus, both a positive value of temperature gradient and motion of continuous solidification front in height of casting may be reached. A positive temperature gradient in casting may be obtained by different methods. All available methods are based on controlling the heat flow in a casting. A direct heat flow is reached at the expense of the application of intense cooling of a mould located below a heating zone (Fig. 1.22) (Blades et al. 2015).

By means of directional solidification the $FeCoCrNiCuTi_{0.8}$ high-entropy alloy was produced at different rates of mould movement (50 μm/s, 100 μm/s, 500 μm/s) (Xu et al. 2020). The results showed that the microstructure was dendritic at all rates of motion.

When the rate of withdrawal has increased the orientation of dendrites becomes homogeneous. Besides, the accumulation of Cr and Ti elements on the solid–liquid interface caused the formation of dendrites. By measuring the initial distance between dendrites (λ_1) and secondary distance between dendrites (λ_2) it was concluded that the structure of dendrites improves evidently with the increase in the rate of withdrawal to 500 μm/s. The maximum compression strength amounted to 1449.8 MPa and maximum hardness 520 HB. Moreover, plastic deformation of the alloy without directional solidification amounted to 2.11%, and that with directional solidification amounted to 12.57% at 500 μm/s. It is proved that directional solidification technology enables one to improve effectively the mechanical properties of $CoCrCuFeNiTi_{0.8}$ high-entropy alloy.

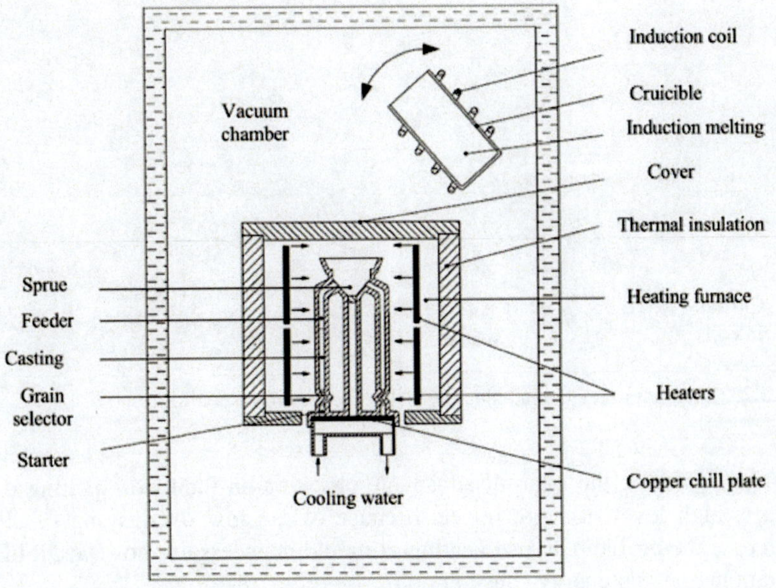

Fig. 1.22 Diagram of directional solidification process (Blades et al. 2015)

Advantages: a large variety of possible microstructures and morphologies may be obtained in changing the values of growth rate and cooling rate. Possibility of formation of monocrystals.

Disadvantages: limitations in the size of produced workpieces.

Levitation Induction Melting

A typical arrangement of setup for levitation induction melting consists of a hollow metal coil with some turns. A zone of low magnetic induction is between a reverse turn of winding in the upper part and levitation water-cooled coils. High-frequency alternating current flows through a coil to produce a variable magnetic field. The magnetic field induces eddy currents in a sample which dissipate energy and produce Lorentz forces supporting a sample in a state of levitation (Fig. 1.23) (Bakhtiyarov and Siginer 2008).

The $HfMo_{0.5}NbTiV_{0.5}Si_x$ high-entropy alloy was produced by means of levitation induction melting in the argon atmosphere (Liu et al. 2017). It was shown that $HfMo_{0.5}NbTiV_{0.5}$ matrix consisted mainly of BCC-phase. Upon adding the Si element, a multi-component silicide $(Hf, Nb, Ti)_5Si_3$ formed inside the alloy. When Si-content amounted to 0.3 the alloy had a hypoeutectic microstructure. With the increase in Si concentration a silicide microstructure underwent an evolution from hypereutectic to eutectic structure and then to hypereutectic one.

Advantages: in comparison with arc melting one of the advantages of levitation induction melting is a possibility of producing large-size samples. The weight of the ingot may amount to 1 kg (Zhang et al. 2020).

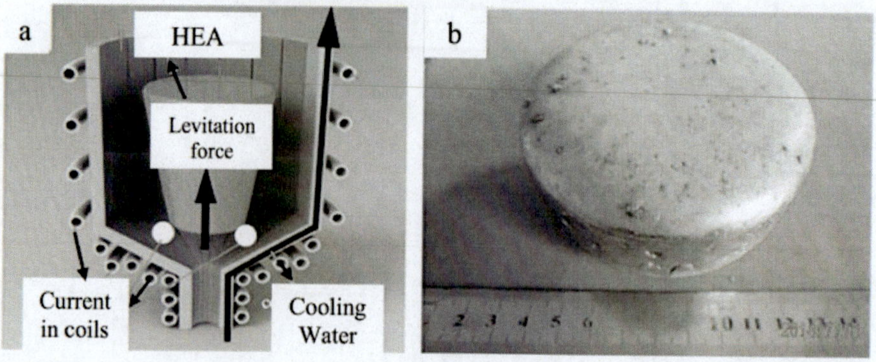

Fig. 1.23 Diagram of levitation induction melting process (Xia et al. 2016)

Disadvantages: the main disadvantage of levitation induction melting is that it gives a high level of ingot rejects because of the low overheating of alloy. To overcome the problem an overheating of mould is necessary; however, it increases contamination and energy consumption.

1.5 Technologies of HEA Coatings Deposition

Along with the production of bulk high-entropy alloys, the problem of deposition of HEA coatings is widely studied nowadays. Laser cladding and magnetron sputtering are the most widespread methods used for the solution of the problem at present time (Fig. 1.24).

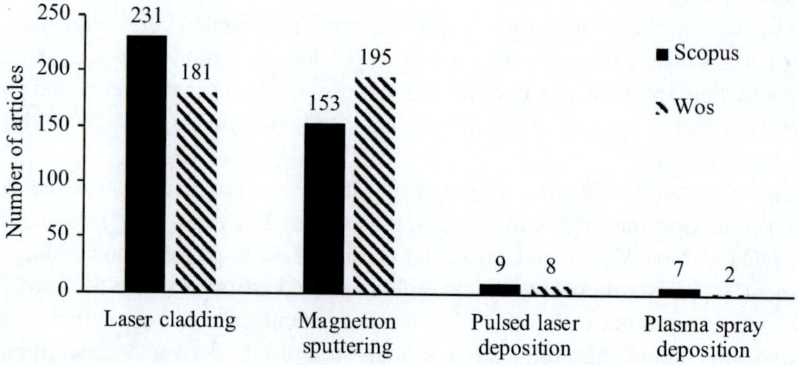

Fig. 1.24 Distribution histogram of the number of works published up to December 2020 depending on methods of HEA coatings deposition

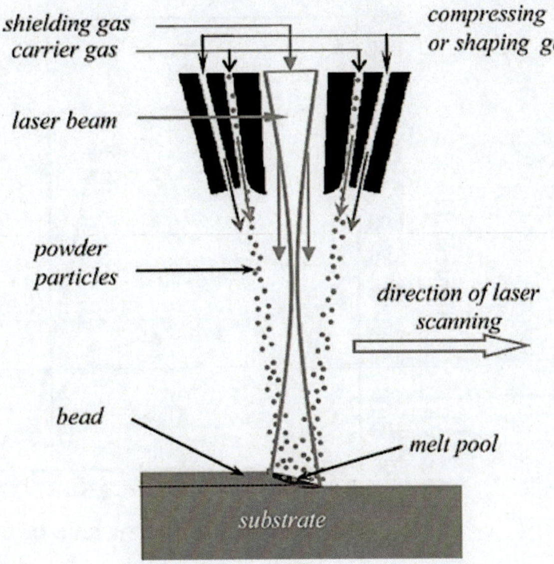

Fig. 1.25 Laser cladding with coaxial powder feeding (Kovalev et al. 2018)

Laser Cladding

Laser cladding is used to produce a relatively thick usually from 50 to 2 mm and homogeneous coating on a substrate. The process is performed by means of a laser beam which melts a substrate surface with simultaneous delivery of the powdered materials into the melt pool as shown in Fig. 1.25 (Kovalev et al. 2018).

In the research (Strife and Passoja 1980) the coatings of $FeNiCoCrTi_{0.5}Nb_x$ alloys ($x = 0.25; 0.50; 0.75; 1.00$) high-entropy alloys on substrate of steel 45 were produced by the laser cladding method. It is shown that the coatings have the following phases: BCC, FCC and Laves phase whose content increases with the increase in Nb in the coating. The hardness of $FeNiCoCrTi_{0.5}Nb_x$ coatings is 2.9 times higher than that of a substrate (Fig. 1.26).

Advantages: the decrease (or elimination) of porosity and cracks as compared to plasma spraying; the decrease in the heat-affected zone and decrease in expenditures for subsequent mechanical treatment in comparison with plasma spraying.

Disadvantages: the high cost of equipment.

Magnetron Sputtering

In magnetron sputtering a high electric field arising from cathode potential accelerates the secondary electrons in the direction normal to the target surface (Fig. 1.27a). Configuration of the magnetic field is usually designed in such a way that it affects the secondary electrons (Simon 2018). Thus, electrons are forced to move along cycloidal drift orbits parallel to the target surface that results in additional atomic ionisation of sprayed gas and higher common plasma currents (Fig. 1.27b). Ions being formed in the drift region have a high hit probability on the cathode situated side by side. It results in a still larger production of secondary electrons and ultimately

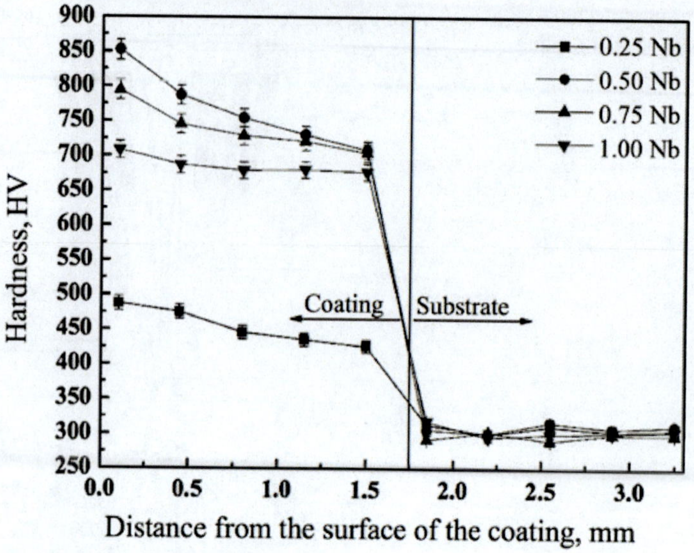

Fig. 1.26 Hardness distribution from the upper part of the coating to a substrate of different coatings $FeNiCoCrTi_{0.5}Nb_x$ ($x = 0.25; 0.50; 0.75; 1.00$) (Zhang et al. 2019)

Fig. 1.27 **a** Magnetic field configuration of the planar magnetron (side view). **b** magnetic field configuration of the planar magnetron (top view) showing EXB orbital drift path of secondary electrons in plasma (Rossnagel 2001)

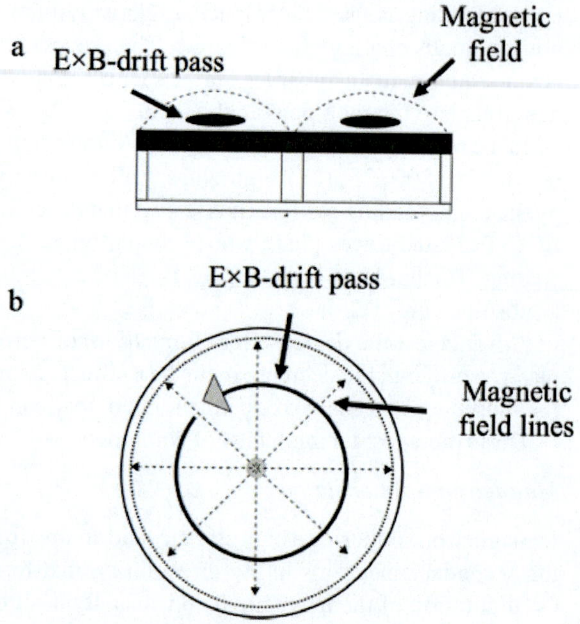

to extremely dense plasma. Due to the magnetic field that is generated from the cathode the ionized atoms affect the target which, in its turn, release particles to cover a substrate.

In the research (Zendejas Medina et al. 2020) the CrMnFeCoNi equimolar alloy was applied by magnetron sputtering method in order to understand the mechanisms of phase formation. It was established that none of the films formed a single-phase structure in the research. Instead, they showed a different amount of FCC-, BCC- and σ-phase as well as MnNi phases and χ phases. It contrasts with the bulk synthesis in which FCC-single-phase crystal lattice is formed. In magnetron sputtering the synthesis takes place below the temperature of stability of single-phase solid solution and a faster diffusion are possible resulting in the formation of equilibrium phases on deposition.

Advantages: high rate of deposition, little increase in substrate temperature on deposition, films are of high purity. The process of sputtering has good reproducibility, and a film of uniform thickness can be produced on a substrate of the large area.

Disadvantages: plasma instabilities. Low coefficient of target use caused by the circular magnetic field generated in magnetron sputtering.

Pulsed Laser Deposition

The pulsed laser deposition process is presented in Fig. 1.28. A beam of pulsed laser is focused on the target material located in a vacuum. A target is usually turned to avoid a repeated ablation from the same spot on the target. The ejected plasma flux impinges a substrate of interest and deposition of a film occurs. As a rule, the growth and quality of film being produced depend on a number of fundamental parameters including substrate selection, substrate temperature as well as absolute and relative kinetic energy and/or rate of feeding the different components to plasma flux. The latter may be affected by the selection of excitation wavelength, duration, energy and intensity of laser pulse, presence (or otherwise) of any background gas and any secondary activation of plasma in a target—substrate gap (Lu et al. 2021).

In the research (Lu et al. 2019) a thin film of $CoCrFeNiAl_{0.3}$ high-entropy alloy was applied on a silicon substrate by means of pulsed laser deposition. The obtained

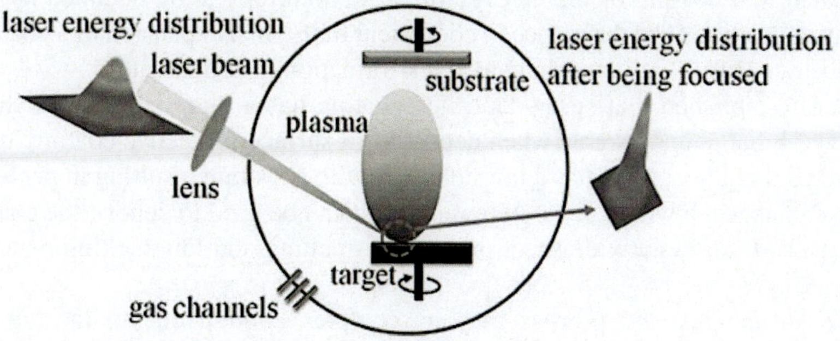

Fig. 1.28 Diagram of pulsed laser deposition process (Lu et al. 2021)

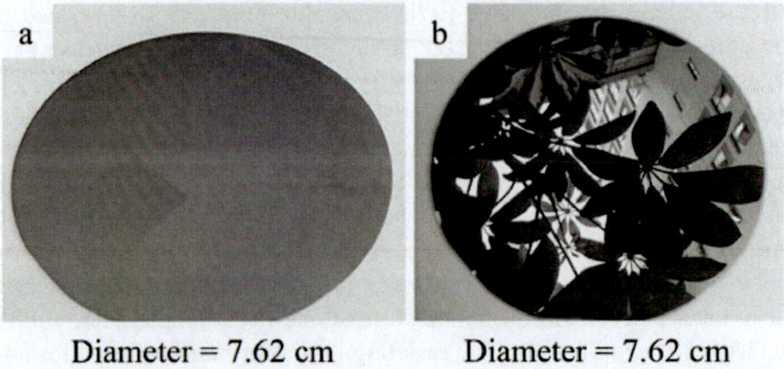

Fig. 1.29 **a** Typical view of $CoCrFeNiAl_{0.3}$ coatings deposited with the use of the method; **b** mirror effect of deposited coatings (Lu et al. 2019)

coating had a simple FCC-structure and showed nanohardness which amounted to 7.66 GPa that approximately three times higher than in bulk HEAs of the same composition. Modulus of elasticity amounted approximately to 150.35 GPa that is lower than in bulk HEAs. The corrosion resistance of obtained coating turned out to be better than in stainless steel 316L. The images of the obtained coatings are presented in Fig. 1.29.

Advantages: the possibility of deposition of very smooth coatings.

Disadvantages: inhomogeneous distribution of energy in laser beam profile may result in inhomogeneous energy profile and angular distribution of energy in laser stub.

Plasma Spray Deposition

Plasma spray deposition is a part of thermal spraying methods in which finely-dispersed metallic and non-metallic materials are deposited in molten or semi-molten state on prepared substrate (Wang 2010). Materials are injected into plasma or plasma jet where particles are accelerated and melted or partially melted before they flatten and solidify on the substrate (Fig. 1.30).

In the research (Wang et al. 2019) a mechanism of solidification cracking was revealed in a coating of the CoCrFeMnNi high-entropy alloy obtained by plasma spray deposition. The difference in coefficient of thermal expansion of a coating and a substrate (Fig. 1.31) contribute to crack formation in solidification.

It is established that spray-deposited coating has a microporosity as shown in Fig. 1.31 that is undesirable when depositing a surface-protective coating. It is also detected that the coating has a low adhesion with substrate resulting in peeling.

Advantages: a wide temperature range that can be used to deposit the coating on materials. It allows a wide spectrum of high-melting and low-melting materials to be used.

Disadvantages: the process may be complex enough due to the amount of interacting parameters. Presence of cracks and peeling.

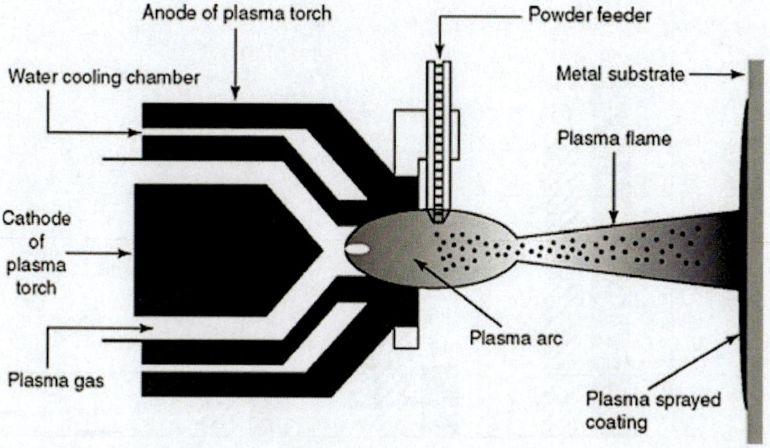

Fig. 1.30 Diagram of plasma spray deposition process (Wang 2010)

Fig. 1.31 Microstructure of air-plasma sprayed layer of CoCrFeMnNi HEA coating (Wang et al. 2019)

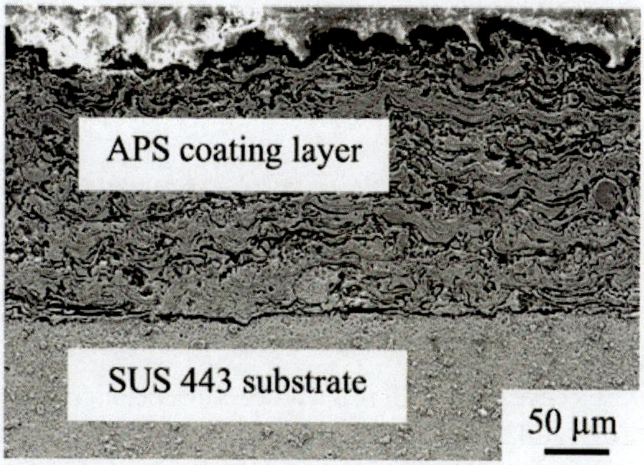

1.6 Technologies of HEAs Additive Manufacturing

Nowadays, technologies of additive manufacturing become more and more popular as methods of high-entropy alloys manufacturing. It is caused by the possibility of producing the products with complex geometry by layer-by-layer forming rather than "subtraction" which is usually performed in conventional melting technologies.

Additive technologies can be divided according to the source of energy that is used to melt the initial material: laser energy, electron beam energy, arc energy and plasma energy. Figure 1.32 presents statistic data on the frequency of use of additive technologies for manufacturing high-entropy alloys. As seen from the figure the largest distribution got the method of selective laser melting. It is followed by plasma arc deposition and direct laser deposition.

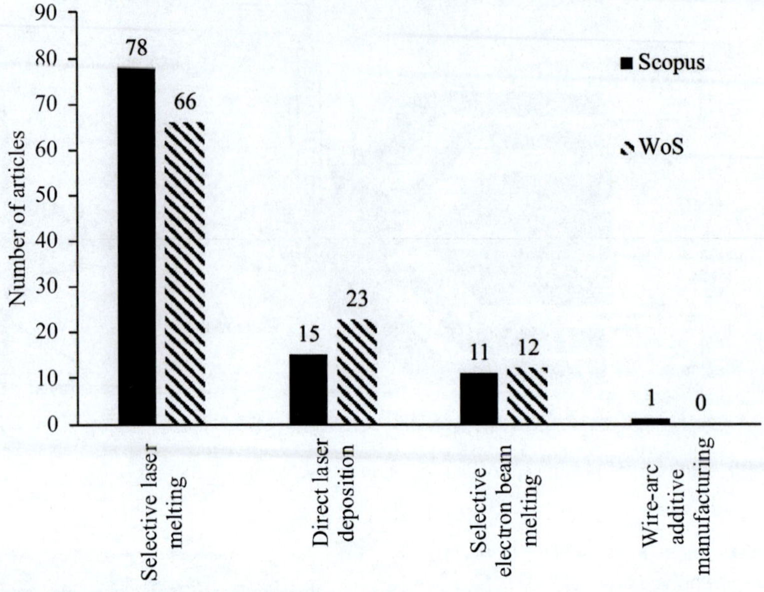

Fig. 1.32 Distribution histogram of the number of works published up to December 2020 depending on technologies of additive manufacturing used to produce high-entropy alloys

Selective electron beam melting, powder plasma arc and wire-arc additive manufacturing finish the list of additive manufacturing methods used for HEA production by the present moment of time.

Selective Laser Melting

Selective laser melting is a process that belongs to the technology of laser powder bed fusion. The method consists of layer-by-layer remelting of powder according to the prescribed 3D model (Fig. 1.33). The powder is bound to the previous layer by local melting by means of the high-intensity laser beam. As shown in Fig. 1.33 the main components of the machine for selective laser melting are laser source, platform on which a part is built, feed container in which powder is kept, and roller for uniform application of powder layer on already hardened layer. Once powder layer has been placed onto base plate, a laser beam selectively scans a layer of powder, monitoring a geometry of a layer. Then a build cylinder is lowered to the value being equal to layer thickness (30–70 μm), roller applies the next layer of powder and process is continued.

In the research (Niu et al. 2019) the AlCoCrFeNi equimolar high-entropy alloy was produced. The effect of bulk density of laser beam energy on the density of the produced product was studied. It is established that with the increase in bulk density of energy the density of product increases linearly as shown in Fig. 1.34.

Advantages: high accuracy of alloying. Possibility of manufacturing the products with complex geometry from refractory materials.

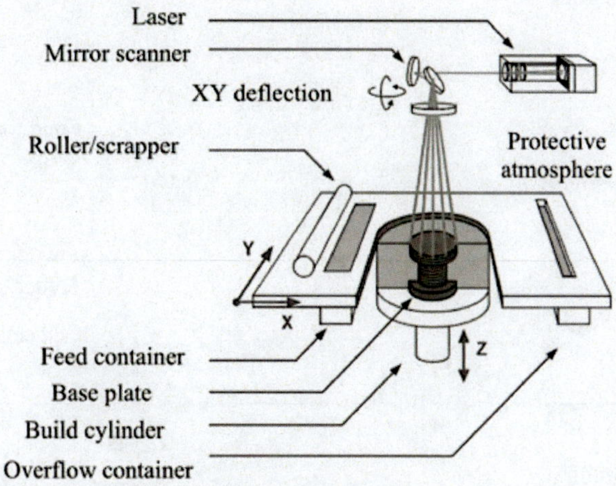

Fig. 1.33 Diagram of selective laser melting process (Kempen et al. 2011)

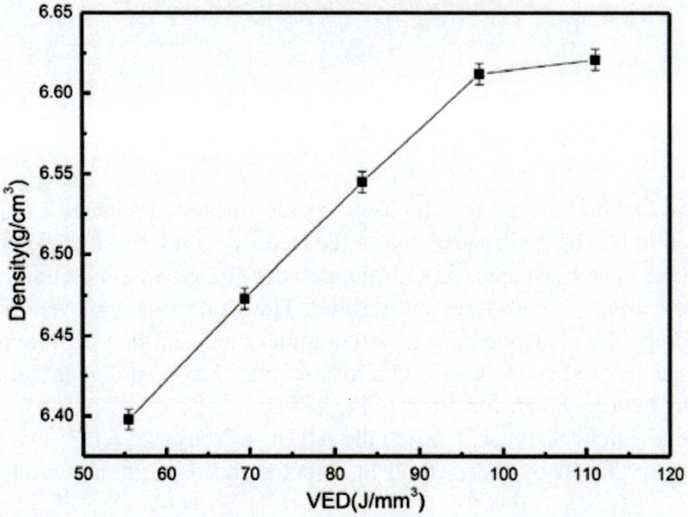

Fig. 1.34 Dependence of bulk density of AlCoCrFeNi equimolar high-entropy alloy on bulk density of laser energy (Niu et al. 2019)

Disadvantages: porosity, presence of non-alloyed particles, the limited size of products due to limited size of the chamber. Considerable stresses are caused by temperature gradients.

Direct Metal Deposition

Direct metal deposition using a powerful laser as a source of energy is the additive manufacturing technology that may be employed for the production of metallic components with high efficiency. The method consists of feeding a powder directly

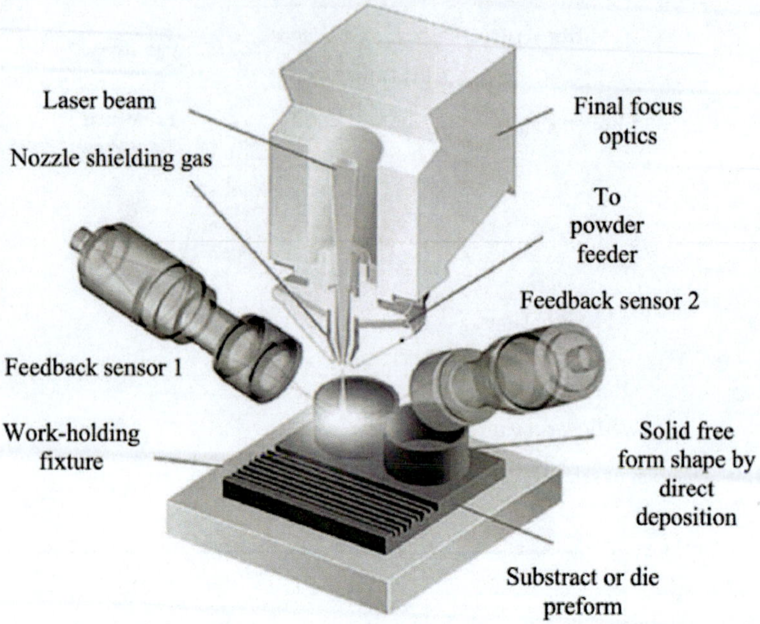

Fig. 1.35 Schematic diagram of direct metal deposition process (Gu et al. 2012)

to the nozzle (Gu et al. 2012). A high-energy laser beam is delivered along the z-axis in the centre of nozzles' matrix and is focused by the lens in a close proximity to a workpiece. The movement of lens and powder nozzles in z-direction controls the height of focuses of both laser and powder. The workpiece is moved in x–y direction by computer-controlled drive system under the beam-powder interaction zone to form the desired geometry of the cross-section. Layers are additively deposited producing a three-dimensional part (Fig. 1.35).

In the research (Sistla et al. 2015) the $Al_xFeCoCrNi_{2-x}$ ($x = 0.3$; 1) high-entropy alloy was produced by the method of direct metal deposition (Fig. 1.36). Phase

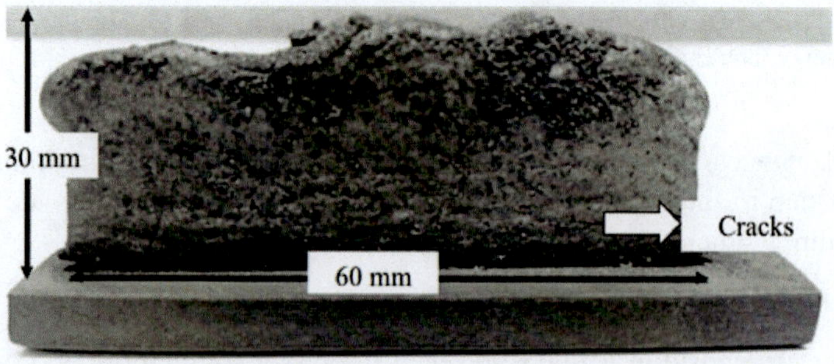

Fig. 1.36 Sample of AlCoCrFeNi equiatomic high-entropy alloy manufactured by direct metal deposition (Sistla et al. 2015)

transformations, microstructure and hardness at different content of Al and Ni both without thermal treatment and after quenching and annealing were studied. It is determined that Al/Ni relation affects deeply the phase transformations, microstructure and hardness of alloys under study. Lattice parameter increased from 0.288 nm to 0.357 nm, and microstructure changed from dendrite to equiaxed and columnar with the decrease in value of Al/Ni relation.

Advantages: cost saving in raw material at the expense of elimination of the production stages necessary for obtaining the pre-alloyed powders. Suitability for fabricating the composition—graduated structures and materials (Gu et al. 2012).

Disadvantages: economical expenses of direct metal deposition system are higher in comparison with other types of additive manufacturing.

Selective Electron Beam Melting

Selective electron beam melting as a version of the method of laser powder bed fusion has recently attracted attention due to its unique characteristics such as high energy density of the incident beam, high rate of scanning and moderate operating expenses. In electron beam melting (Fig. 1.37) a powerful electron beam controlled by electromagnet coils is employed. The electron beam heats the entire layer of powder for each layer of assembly. The whole procedure takes place at high temperature and in a vacuum.

In the research (Shiratori et al. 2016) the AlCoCrFeNi equimolar high-entropy alloy was manufactured (Fig. 1.38) by means of selective electron beam melting. The obtained results were compared with those of the same alloy produced by casting. Both alloys consisted mainly of the fine modulated structure of B2/BCC-phase.

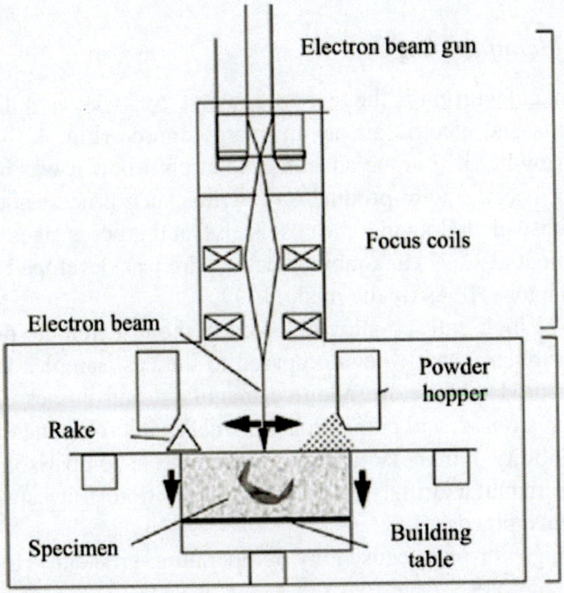

Fig. 1.37 Diagram of selective electron beam melting process (Koike et al. 2011)

Fig. 1.38 The AlCoCrFeNi equiatomic high-entropy alloy samples manufactured by selective electron beam melting (Shiratori et al. 2016)

Though the upper part of the HEA sample had microstructures similar to those of the cast alloy, in the bottom part a roughening of the modulated structure was observed. The FCC-phase precipitated on grain boundaries of B_2/BCC samples. The FCC-phase fraction in the bottom part was much higher than in upper one. Preliminary heating at temperature of 950 °C during manufacturing resulted in FCC-phase formation.

Advantages: large geometrical freedom. Minimum material rejects. Flexible process without expenditures for equipment and adjustment. The elevated temperature of the process reduces to a minimum the residual stresses. Vacuum medium eliminates such impurities as oxides.

Disadvantages: commercially available materials are limited. High roughness of surface.

Wire-Arc Additive Manufacturing

Wire-arc additive manufacturing is the technology that uses wire as initial material for building the parts and electric arc as an energy source (Fig. 1.39). Wire-arc additive manufacturing has its unique advantages but until now it was not used for HEA manufacturing because wire production requires such processes as smelting the raw metal and subsequent drawing, the cost is high and process takes much time. In the research (Shen et al. 2021) a combined cable wire was developed that makes it possible to manufacture HEAs by the method.

The samples of the high-entropy alloy obtained in the research are presented in Fig. 1.40. It was established that when compared to the cast samples the samples produced by the method of wire-arc additive manufacturing possessed a higher ultimate compression strength and elongation but smaller microhardness.

Advantages: possibility of manufacturing large samples in comparison with other methods of additive manufacturing. High rate of deposition. Low cost of initial materials (wire) against powder.

Disadvantages: high stresses caused by temperature gradient. High surface roughness. Presence of pores.

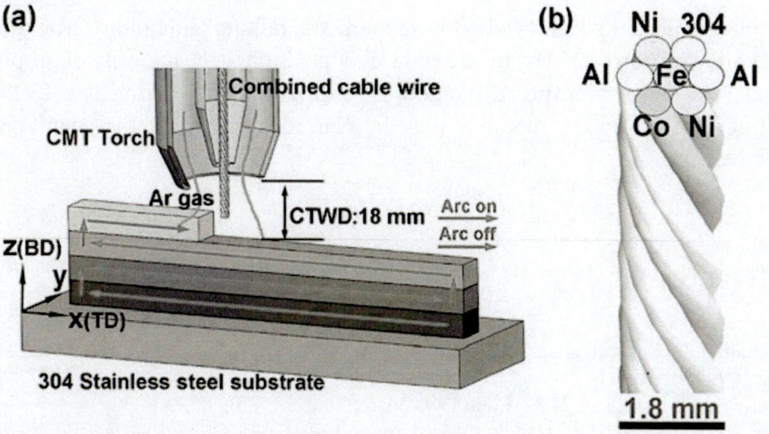

Fig. 1.39 Diagram of wire-arc additive manufacturing process to produce the Al–Co–Cr–Fe–Ni—system high-entropy alloy (Shen et al. 2021)

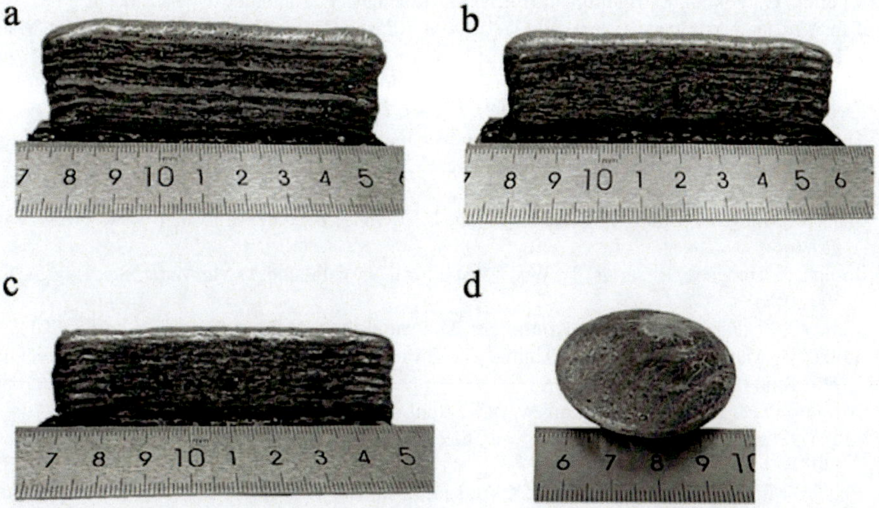

Fig. 1.40 Samples of the Al–Co–Cr–Fe–Ni—system high-entropy alloy produced by wire-arc additive manufacturing (Shen et al. 2021)

1.7 Conclusion

The methods of obtaining high-entropy alloys existing at the present time allow the scope of a large spectrum of possible applications and include both production of bulk materials and deposition of coatings.

Based on the analysis of published works wherein the methods described in the chapter were used it may be concluded that one of the most widespread methods of obtaining bulk samples nowadays is vacuum arc melting technology. Manufacturing the coatings is performed by means of laser deposition in the majority of the presented works. The most widespread method of additive manufacturing is selective laser melting. The choice of methods may be caused by their relative availability and effectiveness.

References

Alshataif, Y.A., Sivasankaran, S., Al-Mufadi, F.A., Alaboodi, A.S., Ammar, H.R.: Met. Mater. Int. **26**, 1099 (2019)

Bakhtiyarov, S.I., Siginer, D.A.: Fluid Dyn. Mater. Process. **4**, 99 (2008)

Blades, A.T., Onyszko, A.: Handb. Cryst. Growth Bulk Cryst. Growth Second Ed., pp. 413–457 (2015).

Cantor, B., Chang, I.T.H., Knight, P., Vincent, A.J.B.: Mater. Sci. Eng. A **375–377**, 213 (2004)

Cavaliere, P.: Spark Plasma Sinter. Mater. Adv. Process. Appl. 1–781 (2019).

Cheng, K.C., Chen, J.H., Stadler, S., Chen, S.H.: Appl. Surf. Sci. **478**, 478 (2019)

Chernik, G., Fokina, E., Budim, N., Hüller, M., Kochnev, V.: Nanoindustry **32** (2007).

Fan, Y., Zhang, Y., Guan, H., Suo, H., He, L.: Xiyou Jinshu Cailiao Yu Gongcheng/Rare Met. Mater. Eng. **42**, 1127 (2013)

Ganesh, U.L., Raghavendra, H.: J. Therm. Anal. Calorim. **139**, 207 (2020)

German, R.M., Suri, P., Park, S.J.: J. Mater. Sci. **44**, 1 (2009)

Gu, D.D., Meiners, W., Wissenbach, K., Poprawe, R.: Int. Mater. Rev. **57**, 133 (2012)

Hu, C., Li, F., Qu, D., Wang, Q., Xie, R., Zhang, H., Peng, S., Bao, Y., Zhou, Y.: Adv. Ceram. Matrix Compos. **164** (2014).

Jian, R., Wang, L., Zhou, S., Zhu, Y., Liang, Y.J., Wang, B., Xue, Y.: Mater. Lett. **278**, 128405 (2020).

Joseph, J., Hodgson, P., Jarvis, T., Wu, X., Stanford, N., Fabijanic, D.M.: Mater. Sci. Eng. A **733**, 59 (2018)

Kempen, K., Yasa, E., Thijs, L., Kruth, J.-P., Van Humbeeck, J.: Phys. Procedia **12**, 255 (2011)

Koike, M., Martinez, K., Guo, L., Chahine, G., Kovacevic, R., Okabe, T.: J. Mater. Process. Technol. **211**, 1400 (2011)

Kovalev, O.B., Bedenko, D.V., Zaitsev, A.V.: Appl. Math. Model. **57**, 339 (2018)

Liu, Y., Zhang, Y., Zhang, H., Wang, N., Chen, X., Zhang, H., Li, Y.: J. Alloys Compd. **694**, 869 (2017)

Lu, T.W., Feng, C.S., Wang, Z., Liao, K.W., Liu, Z.Y., Xie, Y.Z., Hu, J.G., Liao, W.B.: Appl. Surf. Sci. **494**, 72 (2019)

Lu, Y., Huang, G., Wang, S., Mi, C., Wei, S., Tian, F., Li, W., Cao, H., Cheng, Y.: Appl. Surf. Sci. **541**, 148573 (2021).

Mishra, S.K., Pathak, L.C.: Key Eng. Mater. **395**, 15 (2009)

Mohammed, H.G., Albarody, T.M.B., Mustapha, M., Sultan, N.M., Al-Jothery, H.K.M.: Mater. Today Proc. **42**, 2106 (2021)

Nee, A.Y.C.: Handb. Manuf. Eng. Technol. 3487 (2015).

Niu, P.D., Li, R.D., Yuan, T.C., Zhu, S.Y., Chen, C., Wang, M.B., Huang, L.: Intermetallics **104**, 24 (2019)

Rossnagel, S.: In: Handb. Thin Film Depos. Process. Tech., pp. 319–348 (2001)

Sanin, V.N., Yukhvid, V.I., Ikornikov, D.M., Andreev, D.E., Sachkova, N.V., Alymov, M.I.: Dokl. Phys. Chem. **470**, 145 (2016)

Shen, Q., Kong, X., Chen, X.: J. Mater. Sci. Technol. **74**, 136 (2021)

Shiratori, H., Fujieda, T., Yamanaka, K., Koizumi, Y., Kuwabara, K., Kato, T., Chiba, A.: Mater. Sci. Eng. A **656**, 39 (2016)

Shivam, V., Basu, J., Pandey, V.K., Shadangi, Y., Mukhopadhyay, N.K.: Adv. Powder Technol. **29**, 2221 (2018)
Shun, T.-T., Hung, W.-J.: Adv. Mater. Sci. Eng. **2018**, 1 (2018)
Simon, A.H.: In: Handb. Thin Film Depos., Fourth edn, pp. 195–230 Elsevier Inc., (2018).
Sistla, H.R., Newkirk, J.W., Frank Liou, F.: Mater. Des. **81**, 113 (2015).
Strife, J., Passoja, D.: Metall. Trans. **11**, 1341 (1980)
Tan, X., Zhao, R., Ren, B., Zhi, Q., Zhang, G., Liu, Z., Zhao, R., Ren, B., Zhi, Q., Zhang, G., Effects, Z.L., Tan, X., Zhao, R., Ren, B., Zhi, Q., Zhang, G., Liu, Z.: Mater. Sci. Technol. **0836**, 1 (2016)
Torralba, J.M.: In: Hashmi, S., Batalha, G.F., Van Tyne, C.J., Yilbas, B. (eds.) Compr. Mater. Process, pp. 281–294. Elsevier, Oxford (2014)
Wang, M.: In: Ambrosio, L. (eds.) Biomed. Compos., pp. 127–177. Woodhead Publishing, 2010.
Wang, C., Yu, J., Zhang, Y., Yu, Y.: Mater. Des. **182**, 108040 (2019).
Wu, H., Huang, S., Zhao, C., Zhu, H., Xie, Z., Tu, C., Li, X.: Intermetallics **127**, 106983 (2020).
Xia, S., Gao, M.C., Yang, T., Liaw, P.K., Zhang, Y.: J. Nucl. Mater. **480**, 100 (2016)
Xu, Y., Li, C., Huang, Z., Chen, Y., Zhu, L.: Entropy **22**, (2020).
Yeh, J.-W.J.W., Chen, S.K.S.-K., Lin, S.-J.S.J., Gan, J.Y.J.-Y., Chin, T.S.T.-S., Shun, T.-T.T.T., Tsau, C.-H.C.H., Chang, S.-Y.S.Y.: Adv. Eng. Mater. **6**, 299 (2004)
Yim, D., Sathiyamoorthi, P., Hong, S.J., Kim, H.S.: J. Alloys Compd. **781**, 389 (2019)
Zendejas Medina, L., Riekehr, L., Jansson, U.: Surf. Coatings Technol. **403**, 126323 (2020).
Zhang, Y., Xing, Q.: Ref. Modul. Mater. Sci. Mater. Eng., pp. 1–13. Elsevier Ltd., (2020).
Zhang, A., Han, J., Meng, J., Su, B., Li, P.: Mater. Lett. **181**, 82 (2016)
Zhang, Y., Han, T., Xiao, M., Shen, Y.: Optik (Stuttg). **198**, (2019).

Chapter 2
Mechanical Properties and Mechanisms of Deformation of High Entropy Alloys

Abstract In this chapter, the most interesting results and main tendencies on the mechanical properties and mechanisms of deformation of the high-entropy alloys over the last years were analysed. A review of the publications of the last three decades of foreign research on the study of structural-phase states, properties and deformation behaviour of multicomponent high-entropy alloys in different structural states obtained by different methods by methods of modern physical materials science is carried out. The difficulties of comparative analysis and generalisation of data due to different methods of obtaining high-entropy alloys, modes of mechanical tests for uniaxial compression and tension, sizes and shapes of samples, heat treatment, and phase composition are noted. The mechanical behaviour of the different high-entropy alloys systems was considered at low and elevated temperatures. The Research is supported by the Russian Science Foundation (project No. 20-19-00452).

The volume of literary data on the HEA mechanical properties and mechanisms of deformation is extremely great therefore the most interesting results and main tendencies over the last years are presented in short in the section. As it is known the HEA properties are determined by their structure and elemental composition (Pogrebnyak et al. 2014). The high-entropy alloys with BCC lattice have mostly a high strength and low plasticity whereas the materials with FCC—lattice have a low strength and high plasticity.

The researches (Kuznetsov et al. 2012a, b; Gali and George 2013; Ng et al. 2014) are concerned with an investigation into HEA tension-test properties. Thus, in paper (Kuznetsov et al. 2012b) on studying the mechanical properties of AlCrCuNiFeCo high entropy alloy subjected to severe deformation, the improvement of its strength characteristics after high-temperature treatment (abc forging) in comparison with these of the alloy in cast state at room temperature was observed. A brittle-ductile transition is detected in the intervals from 700 to 800 °C for cast alloy and from 600 to 700 °C after the thermal effect. It is worth mentioning that at temperatures above the brittle-ductile transition temperature, a softening of a forged alloy is more noticeable than that of a cast alloy. For example, the ultimate strength decreases from 350 MPa at 700 °C to 180 MPa at 800 °C and to 44 MPa at 1000 °C. A decrease in strength of forged alloy occurs more abruptly: from 350 MPa at 600 °C to 91 MPa

V. E. Gromov et al., *Structure and Properties of High-Entropy Alloys*, Advanced Structured Materials 107, https://doi.org/10.1007/978-3-030-78364-8_2

at 700 °C and 14 MPa at 1000 °C. A brittle-ductile transition occurs at the similar values of ultimate strength (350 MPa) for both states.

In the research (Gali and George 2013), in studying the mechanical properties of CrMnFeCoNi and CrFeCoNi alloys subjected to tension in temperature range from –196 to 1000 °C, their high plasticity is detected at room temperature. Thus, plasticity and hardness increased when temperature decreased, in this case the ultimate strength practically doubled and unit elongation to fracture increased by 1.5 time (to 60%).

Excellent mechanical properties of AlCoCrFeNiTix high entropy alloy at room temperature are described in paper (Zhou et al. 2007). The improvement of mechanical characteristics with regard to large radius of titanium atom were explained by the authors by solid solution strengthening. The best mechanical characteristics are shown by $AlCoCrFeNiTi_{0.5}$ alloy: the yield point is 2.26 GPa the tensile strength is 3.14 GPa, plastic deformation is 23.3%.

A detailed consideration of HEA mechanical characteristics at low temperatures is done in researches (Laktionova et al. 2013; Qiao et al. 2011). The authors of the paper (Qiao et al. 2011), having compared the mechanical properties of AlCoCrFeNi alloy at different temperatures, showed that it possessed excellent mechanical characteristics both at room and cryogenic temperatures. It was detected that the yield point and ultimate strength increased by 29.7 and 19.9% respectively, at temperature decrease from 298 to 77 K. According to the authors' interpretation the mechanical characteristics improve in the presence of a single-phase BCC-structure of solid solution. In that case every atom is considered as "dissolved" in a crystal lattice. As a result of interaction of "dissolved" atoms a crystal lattice deforms and local fields of elastic stresses are formed (Pogrebnyak et al. 2014). Solid solution strengthening results in interaction of dislocation elastic stress fields with analogous fields of "dissolved" atoms which obstructs a motion of atoms and dislocations.

In paper (Hemphill et al. 2012) for $Al_{0.5}CoCrCuFeNi$ high-entropy alloy not only excellent fatigue characteristics were established (fatigue limit varied from 540 to 945 MPa and its ratio to ultimate tensile strength — from 0.402 to 0.703) but also they were compared with characteristics of traditional alloys, e.g. steels, alloys based on Ni, Cu, Mg, bulk metallic glasses. High-entropy alloys can be considered to be perspective materials for use in the fields where the fatigue characteristics are of great practical importance.

The researches (Senkov et al. 2012; Wang et al. 2014) are concerned with studying the mechanical properties of high-entropy alloys at elevated temperatures. The properties of the TaNbHfZrTi alloy are studied in the range from 296 to 1473 K. Three temperature intervals with different mechanisms of deformation were found:

- in the range of 296–873 K strain hardening is temperature independent; there a deformation by twinning and formation of a shear band (673–873 K), high yield point (929 MPa), strong strain hardening (3360 MPa) are observed;
- at 1073 K very small equiaxed grains form along boundaries of deformed grains and twinning deformation is absent;
- in the range of 1273–1473 K a deforming stress decreases abruptly, cracks are absent, the recrystallisation processes proceed during deformation.

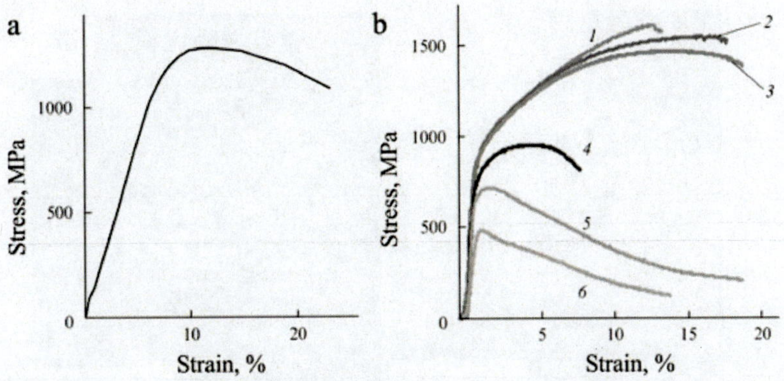

Fig. 2.1 Stress-deformation dependences for the $V_{20}Nb_{20}Mo_{20}Ta_{20}W_{20}$ high-entropy alloy at different temperatures. **a** T = 25 °C; **b** T = 600 (1), 800 (2), 1000 (3), 1200 (4), 1400 (5), 1600 °C (6) (Senkov et al. 2011a)

Table. 2.1 Comparison table of microstructure, lattice parameters and Vickers hardness values of cast high-entropy alloys (Tariq et al. 2013)

Alloys	Microstructure	FCC of constant lattice (Å)	BCC of constant lattice (Å)	Hardness (HV)
AlCoCrCuFeNi	FCC + BCC	3.60	2.87	420
$Al_{0.5}$CoCrCuFeNi	FCC	3.59	–	208
$AlCo_{0.5}$CrCuFeNi	FCC + BCC	3.62	2.87	473
$AlCoCr_{0.5}$CuFeNi	FCC + BCC	3.61	2.87	367
$AlCoCrCu_{0.5}$FeNi	BCC	–	2.87	458
$AlCoCrCuFe_{0.5}$Ni	FCC + BCC	3.61	2.87	418
$AlCoCrCuFeNi_{0.5}$	FCC + BCC	3.63	2.87	423

The HEA based on refractory elements (Senkov et al. 2011a, 2012, 2013; Yang et al. 2012) attracted a special attention of researchers. The analysis of stress–strain dependences for the $V_{20}Nb_{20}Ta_{20}W_{20}$ high-entropy alloy at different temperatures (Fig. 2.1) showed that it had a high yield point (1246 MPa) and limited plasticity (1.5%) at a room temperature. In the range from a room temperature to 600 °C the alloy undergoes a brittle-ductile transition which is accompanied by decrease in strength. With a further temperature increase the yield point decreases from 1246 to 477 MPa. In a temperature range of 1400–1600 °C a softening of the alloy takes place. In refractory high-entropy alloys a resistance to high-temperature softening (Pogrebnyak et al. 2014) is strong due to a weak diffusion of elements.

In Tung et al. (2007) the phase composition and hardness of the AlCoCr-CuFeNi alloys (Table 2.1) were determined. It turned out that hardness of the $Al_{0.5}$CoCrCuFeNi alloy was far less than in other alloys. It may be connected with the fact that there is much more FCC-phase in the material while BCC-phase predominated in other ones. The largest value of hardness was observed in the

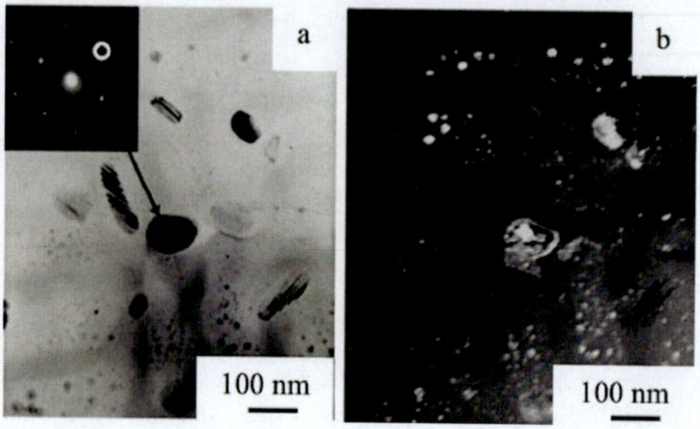

Fig. 2.2 Microstructures of AlCoCrCuFeNi cast alloy, **a** in a bright field, **b** in a dark field (Yeh et al. 2007)

$AlCoCrCu_{0.5}FeNi$ alloy where the amount of copper is the least. It was explained by the decrease in a fraction of interdendritic areas and the increase in amount of BCC-phase dendrites (Tung et al. 2007; Ivchenko 2015).

It is important to note that a tendency to form the simple (FCC, BCC, HCP) solid solution multi-component phases (Ivchenko 2015) is not only increased in high-entropy alloys but also the formation of nanodimensional phases is observed. Figure 2.2 demonstrates a microstructure of the AlCoCrCuFeNi cast equimolar alloy (Yeh et al. 2007). In electron micrographs, especially in a dark-field ones, a large quantity of nanoprecipitations are seen inside the grains matrix. In fact, the nanophases were observed in a matrix of high-entropy cast alloys if the corresponding methods of the high resolution were used and first of all, the transmission electron microscopy (TEM) (Ivchenko 2015).

The formation of nanodimensional precipitations may be connected with the deceleration of decay kinetics. In the process of diffusion-controlled transformations the new phases requiring a joint diffusion of different atoms are formed in the oversaturated matrix by nucleation and growth. It may be supposed that in high-entropy alloys the decay of substitution elements of oversaturated multi-component solutions will occur firstly, in the conditions of formation of a very large quantity of nuclei, and secondly, in the conditions of their strong competition to subsequent growth or, in other words, retarded kinetics that, on the whole, will result in the formation of multi-element nanophases (Ivchenko 2015).

Next, it is worth paying attention to the research (Kuznetsov et al. 2012a) wherein the effect of microstructure on mechanical properties under the tension of AlCoCr-CuFeNi high-entropy alloy was studied. The samples of the alloy's ingot were subjected to abc (all-side) hot isothermic deformation by pressing.

The treatment was connected with the decrease in temperature of brittle-ductile transition from the interval of 700–800 °C in a cast state to 600–700 °C after abc-deformation. At room temperature a deformed alloy is stronger and more plastic than the cast one: the ultimate strength in a cast state amounted to 790 MPa, and after abc-pressing it was 1170 MPa. At temperatures above the brittle-ductile transition, a considerable decrease in flow stress took place. The ultimate strength of a cast alloy decreased from 350 MPa at 700 °C to 180 MPa at 800 °C and to 44 MPa at 1000 °C. The decrease in the ultimate strength of a deformed alloy occurred still quicker: the ultimate strength decreased from 350 MPa at 600 to 91 MPa at 700 °C and to 14 MPa at 1000 °C. The reduction of strength was accompanied by a marked increase in plasticity. In a temperature interval of 800–1000 °C the deformed alloy demonstrated a superplastic behaviour. The unit elongation exceeded 400% and at a temperature of 1000 °C, it amounted to 860% (Kuznetsov et al. 2012a).

The study of thermomechanical treatment effect by the method of hot rolling on the structure and mechanical properties of alloys based on simple solid solutions by the example of the CoCrFeNi and CoCrFeNiMn alloys was considered in the research (Gali and George 2013). Prior to rolling the alloys were subjected to homogenisation at 1000 °C for 24 h in vacuum for improving the homogeneity in chemical composition. The thermomechanical treatment was performed at a temperature of 1000 °C. In order to prevent the heat losses during rolling the samples were welded into the shell of stainless steel sheets. Prior to every pass, the samples were preheated in the furnace for 15 min at a temperature of 1000 °C. The total degree of alloys deformation amounted to 92%. As a result of the treatment the completely recrystallized microstructure of CoCrFeNi and CoCrFeNiMn alloys with the grain size of 32 and 11 μm, respectively, was obtained. To relieve residual stress, the final annealing at a temperature of 900 °C for 1 h was carried out. The investigations into mechanical properties showed that the alloys had a low yield point and high values of strength and plasticity. The ultimate strength of CoCrFeNi alloy at a room temperature and rate of deformation of 10^{-3} s^{-1} amounted to ≈600 MPa, the yield point was 210 MPa, the unit elongation was of the order 40%. The addition of Mn to the alloy increases slightly the strength properties: $\sigma_b = 670$ MPa, $\sigma_{0.2} = 210$ MPa and $\delta = 45\%$. The decrease in test temperature results in the growth of strength and plastic qualities. A possible reason for high degree of strengthening at low temperatures of deformation is twinning (Shaisultanov 2015).

In addition to high strength properties at cryogenic temperatures, the CoCrFeMnNi five-component alloy demonstrates the high values of impact strength (Gludovatz et al. 1153). The alloy was produced by arc melting and casting into the water-cooled copper mould of rectangular cross-section and then it was treated by cold forging and rolling at room temperature. The thickness of sheets amounted to 10 mm. After treatment, the alloy had an equiaxed grain structure. In the course of tests, the impact strength values exceeding $K = 300$ MPa $m^{1/2}$ ($J = 500$ kJ/m^2) at 77 K were obtained. The high values of impact strength differ favourably from the high alloy austenitic stainless steels such as 304L and 316L (Mills 1997) that have strength in the range of $K_Q =$ from 175 to 400 MPa $m^{1/2}$ at room temperature and the best cryogenic steels such as 5Ni or 9Ni with $K_Q =$ from 100 to 325 MPa $m^{1/2}$ at 77 K (Strife and Passoja 1980).

In the research (George et al. 2020) the most detailed analysis of HEA mechanical properties and deformation mechanisms as of the end of 2018 is presented. The authors carried out an exhaustive comparison with the data on composite concentrated alloys (CCA) having analysed more than 180 papers (Ng et al. 2014; Wang et al. 2007, 2015, 2016a, b, c, d, 2017a, b, 2018; Dirras et al. 2015, 2016a, b; Daoud et al. 2015; Gludovatz et al. 2015; Lee et al. 2015, 2016, 2017, 2018; He et al. 2014, 2016a, b, 2017a, b; Mridha et al. 2015; Fu et al. 2015, 2016, 2014, 2017, 2018; Park et al. 2015; Schuh et al. 2015, 2018; Guo et al. 2015, 2017; Jiao et al. 2015a, b, 2016; Kourov et al. 2015; Ji 2015; Zhang et al. 2015, 2016a, b, c, 2017a, b, c, d; Ma et al. 2015, 2016, 2017, 2018; Kumar et al. 2015; Chang and Yeh 2015; Stepanov et al. 2015a, b, 2017; Jiang et al. 2015a, b, 2016a, b; Chen et al. 2015, 2016a, b, 2017, 2018; Zaddach et al. 2013; Senkov and Semiatin 2015; Li et al. 2016a, b, c, d, 2017a, b, c, d, 2018; Yang et al. 2015, 2017; Pickering et al. 2016; Maiti and Steurer 2016; Yu et al. 2016; Rogal et al. 2016; Komarasamy et al. 2015; Wani et al. 2016; Munitz et al. 2016; Okamoto et al. 2016; Juan et al. 2016a, b; Bardziński and Błyskun 2016; Niu et al. 2016; Pogrebnjak et al. 2014; Yurkova et al. 2016; Xiao et al. 2017; Moravcik et al. 2016, 2017; Tian et al. 2016, 2017; Jin et al. 2016, 2017; Jang et al. 2016; Yurchenko et al. 2016, 2017; Pi et al. 2016; Salishchev et al. 2014; Shahmir et al. 2017; Lee and Shun 2016; Tazuddin and Gurao 2016; Wu et al. 2014, 2017a, b; Reddy et al. 2017; Cakmak 2017; Maier-Kiener et al. 2017; Prasad et al. 2017; Gwalani et al. 2017, 2018; Xian et al. 2017; Liang et al. 2017; Rao et al. 2017; Tsao et al. 2017a, b; Wang and Xu 2017; Joo et al. 2017; Tabachnikova et al. 2017; Gong et al. 2017; Yao et al. 2017; Hong et al. 2017; Braeckman et al. 2017; Raghavan et al. 2017; Joseph et al. 2015, 2017a, b; Liu et al. 2015a, b, 2017a, b, c; Varvenne and Curtin 2017, 2018; Lin et al. 2015; Luo et al. 2017; Ogura et al. 2017; Ahmad et al. 2017; Feng et al. 2017a, b, 2018; Shaysultanov et al. 2017a, b; Ming et al. 2017; Vida et al. 2017; Kawasaki et al. 2017; Huo et al. 2017a, b; Lim et al. 2017; Sun et al. 2017; Sathiyamoorthi et al. 2017; Cai et al. 2017; Niendorf et al. 2018; Wei et al. 2018; Meng et al. 2015; Xu et al. 2018; Seol et al. 2018; Bönisch et al. 2018; Lei et al. 2018; Zhu et al. 2017; Haase and Barrales-Mora 2018; Guo 2015; Laplanche et al. 2018a; Slone et al. 2018; Sosa et al. 2015; Choudhuri et al. 2015). Figure 2.3 represents the mechanical properties under tension and compression. The

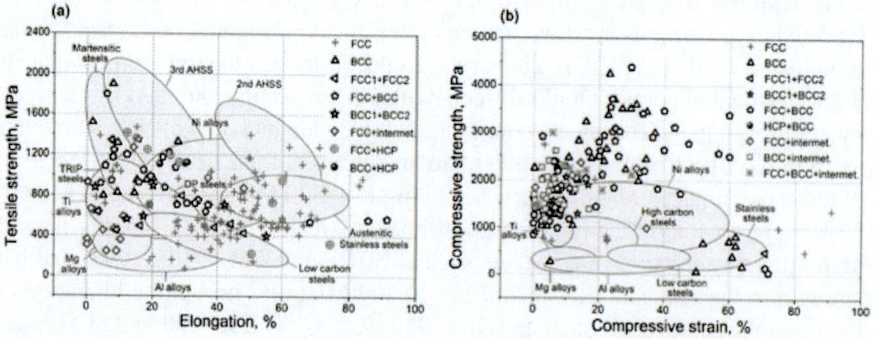

Fig. 2.3 Ultimate tensile strength at room temperature depending on elongation to fracture (**a**) and compressive strength depending on compressive strain (**b**) of HEA and CCA described in literature

refinements and additional information on the dependences are given in https://doi.org/10.1016/j.actamat.2019.12.015. The authors George et al. (2020) make some general conclusions from Fig. 2.3a, Firstly, it is difficult to consider that HEA and CCA demonstrate the strength and plastic properties exceeding the classical structural alloys (martensite steels, high-strength steels, Ni-based alloys, etc.). Secondly, the HEA and CCA encompass totally a spectrum of properties of steels, alloys based on Al, Ti, Mg, Ni. Thirdly, HEA and CCA occupied entirely the regions both with high plasticity and low strength and vice versa (George et al. 2020).

See Appendix Table 1 for detailed information on data sources. The 2nd and 3rd AHSS designate two generations of modern high-strength steels, DP steels for two-phase steels and TRIP steel for steels with plastic transformation (George et al. 2020).

The characteristics under compression (Fig. 2.3b) exhibit another tendency in comparison with the data on tension. The markedly higher properties compared with structural alloys are reached, for example, in $Al_{0.3}CoCrFeNi$ high-entropy alloy (the ultimate strength 1378 MPa, deformation to fracture 97%). The experimental complications permit no comparison of the results in the same HEAs both under compression and tension (George et al. 2020). The situation is aggravated by the fact that HEAs of the same composition may have a FCC, BCC structure or be two-phase ones; the grain size is a decisive factor as well (George et al. 2020).

We begin the analysis of deformation mechanisms with the HEAs with FCC structure because data on them are larger in literature. Unlike pure FCC metals the Cantor alloys (CrMnFeCoNi) demonstrate a high-temperature dependence of yield point (Fig. 2.4) in the range of 77–500 K with different grain sizes (Koshelev 1971). The authors of the papers (Koshelev 1971; Haglund et al. 2015; Laplanche et al. 2015, 2018b) give a negative answer to arising evident question concerning the relation of

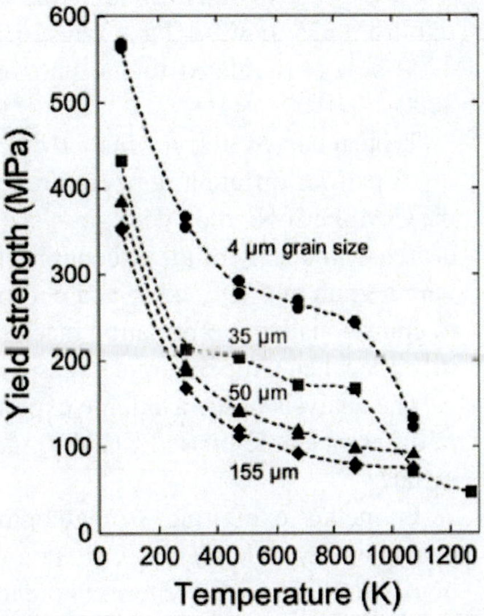

Fig. 2.4 Dependence of yield point of CrMnFeCoNi high-entropy alloy on temperature and grain size. Adapted from Gali and George (2013)

the temperature dependence of yield point σ_T (T) with the abnormal temperature dependence of shear modulus because the dependence σ_T (T) retains even after shear modulus normalisation.

The frequently encountered statements that HEAs should be stronger than common alloys with less quantity of elements (Yeh 2015) are inconsistent with the conclusions from Fig. 2.3. In this comparison, it is necessary for the alloys to have a similar crystal structure, grain size, dislocation density, etc. The increase in the number of elements from four in CrFeCoNi to five in CrMnFeCoNi left no substantial change in yield point (Gali and George 2013). The CrCoNi ternary alloy has a higher strength than Cantor five-component alloy CrMnFeCoNi and all four-component alloys FeNiCoCr, NiCoCrMn and FeNiCoMn. The alloys with the identical number of elements had a noticeably different strength: the FeNiCoCr was far stronger than FeNiCoMn, the CrNiCo was stronger than MnFeNi. That is the strength is not determined by the number of alloying elements and their type. In the authors' opinion (Varvenne et al. 2016) the decisive importance has a volume inconsistency of the alloying element. The atoms with larger effective radii have a tendency to considerably displace smaller atoms and the latter tend to distribute between the large atoms in order to adapt to displacements (George et al. 2020). The values of activation volume for Cantor alloys obtained from the data on stress relaxation amount to $\approx 60b^3$ 77 K and $\approx 360b^3$ at room temperature which exceed the corresponding values of point defects ($\sim b^3$) but is less for solid solutions (George et al. 2020). When analysing the effect of deformation rate the authors (Gali and George 2013) come to the conclusion that in quasi-static tests (10^{-1}–10^{-5} s^{-1}) for alloy with grain size 35 μm the effect of deformation rate is negligible (George et al. 2020). In the range of 10^{-3}–10^{2} s^{-1} the yield point increases about by 35% from 484 to 650 MPa for the alloy with the grain size of 10 μm with the corresponding increase in ultimate strength from 853 to 968 MPa.

At large rates of deformation, the effect of deformation rate is still more substantial: from 325 to 360 MPa at rates of 10^{-4}–10^{-2} s^{-1} to 590–680 MPa at rates 3000–4700 s^{-1}. It is related to the phonon mechanism of deceleration (George et al. 2020).

Typical curves $\sigma(\varepsilon)$ for the CrMnFeCoNi high-entropy alloy with the grain size of 50 μm for different temperatures are shown in Fig. 2.5. Under compression of the CrMnFeCoNi alloy with grain size of 5–150 μm the flow stresses grow with the decrease in size with the exception of high temperatures (800 °C) where the alloy with a grain of 5 μm exhibits a deformation softening after yielding primarily due to grain-boundary deformation mechanisms. The yield point, flow stress, elongation increase when temperature decreases up to temperature of liquid nitrogen.

The analysis of the available experimental results concerning the yielding points of different single-phase FCC alloys and HEAs is indicative of similarity with FCC metals.

In metals exhibiting strengthening only due to dislocation interaction (for example, Taylor hardening) the rate of strain hardening usually decreases continuously until Consider criterion is reached and instability of neck formation occurs followed by failure in short time. The TEM study of samples taken from tension tests

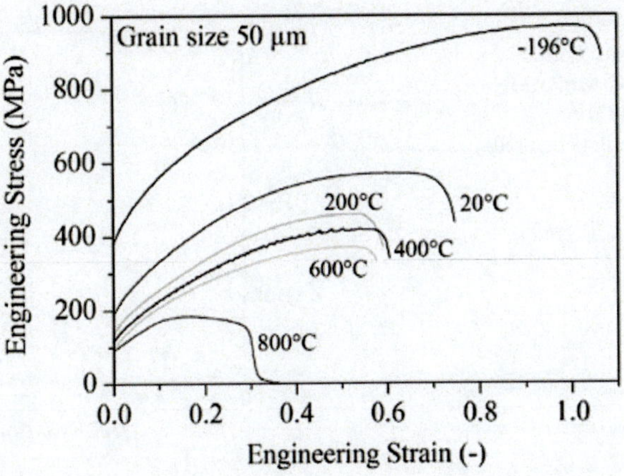

Fig. 2.5 Stress–strain curves for CrMnFeCoNi high-entropy alloy at different temperatures (George et al. 2020)

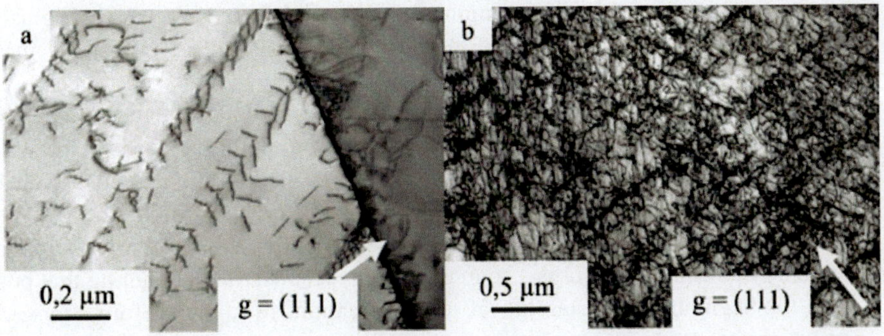

Fig. 2.6 Planar rows of dislocations on planes {111} in CrMnFeCoNi high-entropy alloy under low deformations, ~5% deformation at 77 K (at the left) forming cellular-net-like and ball structures under high deformations; ~22% deformation at room temperature (at the right) (Laplanche et al. 2016)

interrupted after different degrees of deformation (Otto et al. 2013; Laplanche et al. 2016) showed that dislocation substructure develops from a plane sliding of chaotic dislocations under small deformations to transversely sliding ball dislocations under intermediate deformations and cellular structures under high deformations (Fig. 2.6) likewise in microstructural evolution being observed in the majority of FCC metals. The increase in dislocation density with deformation was measured by means of TEM, and the results are shown in Fig. 2.7. As illustrated in the figure the increase in dislocation density 77 K is practically the same as at room temperature. It means that if Taylor strengthening is the only acting mechanism, the rate of strain hardening normalized in shear modulus as a function of deformation should be identical at both temperatures (because it depends only on the square root of dislocations' density and a pair of other constants being deformation-independent). However, the strain

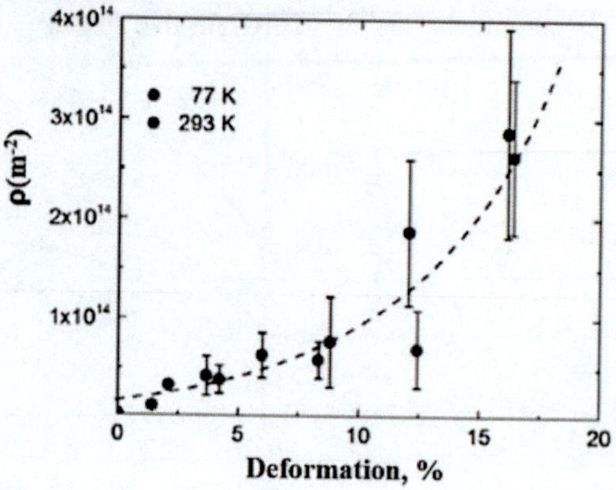

Fig. 2.7 Increase in dislocation density with deformation in CrMnFeCoNi alloy at 77 K and room temperature (Laplanche et al. 2016)

hardening rates at room temperature and 77 K differ noticeably with explicit discrepancy beyond ~6% plastic deformation. Therefore, in addition to simple dislocation hardening at 77 K another mechanism should work.

The mechanism involves a twinning caused by deformation which begins to display themselves in some grains after ≈6% deformation and constantly in all investigated grains after ≈9% deformation. The twins in nanometer scale represent the normal twins being observed in FCC-metals and they contribute to strengthening as earlier reported in other papers. With an increase in deformation their thickness and volume fraction increase (Laplanche et al. 2016). It was suggested (Meyers et al. 2001; Kireeva et al. 2016, 2018) that mechanical twins promote strengthening dividing grains into still finer grains (so-called Hall–Petch effect). The twinning caused by deformation ensures a source of highly stable strain hardening that postpones the beginning of instability of neck formation till higher deformations. As a result, the twinning effect can simultaneously increase in strength and plasticity. Some twin boundaries act as barriers for the motion of dislocations while the other enable partial dislocations to slide along the twin-matrix interface apparently thus relieving some accumulated stresses and ensuring further deformation.

It is believed that in FCC-metals there is a critical stress for twinning that depends weakly, if at all, on temperature. Supposing that it takes place in Cantor alloy, it might be reasonable to expect a twinning at room temperature if internal stress can grow considerably. Axial stress at which twins appear at 77 K is ~720 MPa (Laplanche et al. 2016). At room temperature, such a high stress is reached only in the immediate vicinity of failure. Another factor that, as it is known, affects twinning is the grain size: a probability of twinning caused by deformation decreases as the grain size is decreased (Meyers et al. 2001).

The majority of papers on monocrystalline alloys testify to the decisive role of monocrystals' orientation under tests (Meyers et al. 2001; Kireeva et al. 2016, 2018; Ma et al. 2014; Yasuda et al. 2015). In this case, it is important what precise methods to determine the twinning were used. The TEM methods enable one to register the beginning of twinning at $\varepsilon > 5\%$ of samples oriented for multiple slips while the main role belongs to dislocation activity at orientation for single sliding initially to 27%.

At orientation of multiple sliding [001] multiple slip orientation [001] (Osintsev et al. 2021) the twinning was not observed at all. It is also known that twinning is impeded with the decrease in grain size, it is realized under lower stresses in monocrystals. It is necessary to note that diffraction of inversely dissipated electrons and other methods rank below TEM in assessing the role of twinning (George et al. 2020).

BBC-based HEAs

In the research (Qi and Chrzan 2014) it was suggested that the plasticity of BCC-alloys can be controlled by changing in a valence electron concentration (VEC). Senkov et al. developed (Senkov et al. 2010) MoNbTaV and MoNbTaVW alloys with high yield point and low plasticity.

In comparison with FCC-alloys discussed above, there is far less information on the detailed deformation behaviour of BCC-high-entropy alloys probably because the majority of them are brittle at low homologous temperatures. Thus, from several studied single-phase BCC HEAs including NbMoTaW, VNbMoTaW, TiVZrNbHf, and TiZrNbHfTa (Senkov et al. 2011b) only TiZrNbHfTa is known as plastic under tension. The alloy is very ductile and can be subjected to severe rolling at room temperature (thickness decrease >80%). Consequently, its cast structure can be fractured, and fully recrystallized microstructure can be systematically studied depending on grain size. Since the first report on tensile properties in 2015, it has been exciting an ever-increasing interest and nowadays it is the most studied BCC-HEA from the point of view of principal mechanisms.

After arc melting and pressing (1473 K, 207 MPa, 3 h) a dendritic structure disappearing after thermo-mechanical treatment (cold rolling + recrystallisation for 2 h at 1000 °C) is formed to constitute a homogeneous grain structure and chemical composition (Fig. 2.8).

In comparison with FCC—HEA (for instance, CrMnFeCoNi), BCC—alloy TiZrNbHfTa strengthens less. The estimation of α—parameter in the expression $\Delta\tau = \alpha G b \sqrt{\rho}$ is indicative of the decisive role of Peierls barrier. For CrMnFeCoNi alloy with FCC—structure the main contribution is made by "forest" dislocations.

TEM analysis enables the authors to make a conclusion on decisive Taylor mechanism of strengthening (George et al. 2020).

Figure 2.9 shows the curves of true stress–strain under tension of TiZrNbHfTa high-entropy alloy in three conditions (Li et al. 2016a). In as-rolled state the alloy showed an intermediate strength with limited plasticity and work softening soon after yielding. After annealing at 800 °C the yield point increased but was accompanied by a marked decrease in plasticity. Analysis of microstructure showed that after the treatment a partial recrystallisation occurred but the initial single-phase BCC—structure decomposed into two BCC—phases with different lattice parameters (Li et al. 2016a). After annealing at 1000 °C the microstructure recrystallized fully and consisted of one BCC—phase with the best combination of mechanical properties: yield point 1145 MPa, ultimate tensile strength 1262 MPa and ~8% unit elongation to failure. The surface of fracture had both plastic (transcrystallite) and brittle (intercrystallite) regions. Senkov and Semyatin (Li et al. 2016a) proposed that brittle films

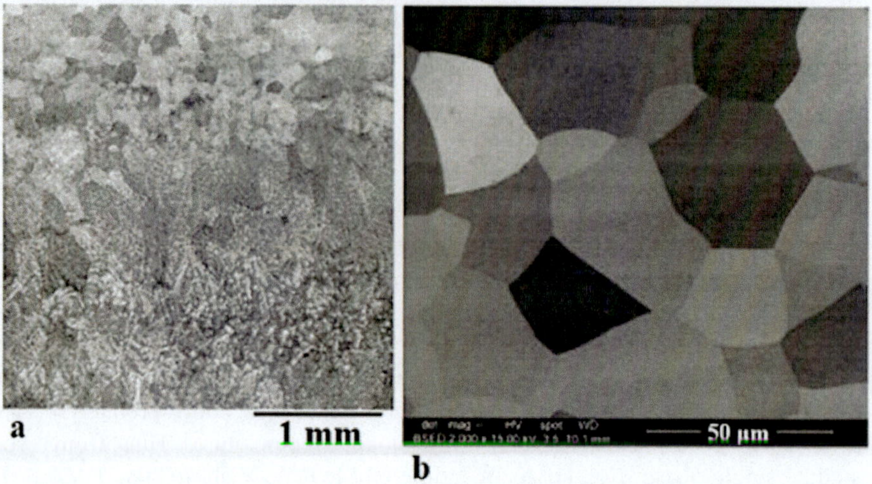

Fig. 2.8 Microstructure of TiZrNbHfTa HEA after casting and pressing (**a**). **b** SEM image in inversely reflected electrons after cold rolling with 65% degree of compression and annealing for 2 h (Li et al. 2016a; Kilmametov et al. 2019)

Fig. 2.9 True strain curves of TiZrNbHfTa high-entropy alloy in three different microstructural states (Li et al. 2016a)

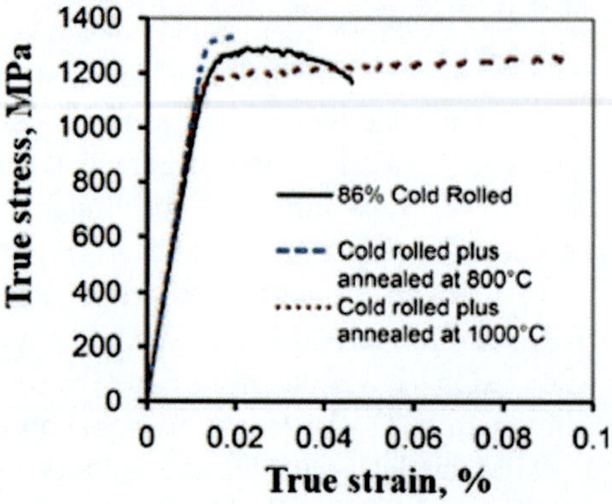

of the second phase could be present on grain boundaries that, probably, contributed to intergranular fracture being observed.

The effect of grain size on tension—test properties of TiZrNbHfTa HEA was studied in a later research and it was found that yield point was approximately the same (~950 MPa) for grain sizes of 38, 81 and 128 μm (Juan et al. 2016a). The value is significantly lower than the yield point (1145 MPa) mentioned above which was for grain size of 22 μm (Li et al. 2016a) and it cannot be explained as Hall–Petch effect.

Activation volumes at room temperature were measured using the load relaxation method and amounted to ~40–50b^3 at stage I of strain hardening (up to ~1% plastic deformation) and decreased to ~30b^3 at the end of stage II (~10% deformation) and III (Munitz et al. 2016). The authors came to the conclusion that the low values are identical to those that were measured in other BCC-metals and alloys and are indicative of a strong Peierls barrier. Besides, a weak dependence of apparent activation volume on plastic deformation suggests that hardening makes only a moderate contribution to flow stress that is consistent with low value of Taylor hardening parameter (α).

In overview (George et al. 2020) it is noted that improvement of HEA properties with a BCC lattice can be achieved through the implementation of other hardening mechanisms (metastable solutions, short-range hardening, etc.). Thus, ordered oxygen nanoscale clusters in TiZrHfNb HEA are capable of fixing dislocations, stimulating their multiplication, and changing the strength properties.

Summarising the results of the analysis of mechanical properties and deformation mechanisms the authors (George et al. 2020) believe that, fortunately, HEA deformation is apparently controlled by ordinary phenomena connected with dislocations. Therefore, complications arise in understanding of the fact how dislocations move through a lattice severely disordered in composition. Advance in understanding a flow stress, achieved up to the day, is indicative of a more general theoretical approach to studying/predicting how spatial fluctuations of dissolved substances distribution in different scales are connected with behaviour of crystal defects, in the first place, such as dislocations, vacancies, grain boundaries and cracks. The behaviour of the defects determines a diffusion transfer, grain—boundary strengthening, grain growth, failure and other important phenomena. However, because quantitative theories of mechanical properties such as strain hardening, plasticity, twinning, failure and fatigue do not exist for traditional technical alloys including dilute alloys based on solid solutions, a development of analogous theories in very complex HEAs remains a serious problem. Experimental results demonstrating high characteristics of HEAs are a strong motive force for solving the task.

The author (Rogachev 2020) analysed the mechanical properties of HEAs for the last publications of 2018–2020. It is noted that good mechanical properties are inherent not only in cast HEAs but alloys produced by other ways. Thus, Cantor alloy CoCrFeNiMn, synthesized by the method of severe plastic torsional strain under pressure, showed a record hardness of 6.7 GPa for the alloys (possibly, due to a nanocrystalline structure and inclusions of chromium oxide) (Kilmametov et al. 2019).

In recent years the HEAs, consolidated on additive technology of selective laser melting (SLM), have been developing. Initial powders for SLM of required round shape are produced by gas atomisation or crushing followed by plasma spheroidisation of cast HEAs. The materials manufactured by the SLM method combine a good strength (about 1000 MPa under tension) with a high plasticity (~10–30% ultimate breaking deformation) (Rogachev 2020; Niu et al. 2019; Yao et al. 2020; Yang et al. 2019a, b). Microhardness of SLM—alloy CoCrFeNiAl amounted to 632.8 HV.

One of the most attractive features of HEAs is a high strength and plasticity at lower and even cryogenic temperatures (Murty et al. 2019; Zhang 2019). Some results are presented in Table 2.2. It is also reported about high fracture toughness of HEA at low temperatures, for example, 232 MPa·m$^{1/2}$ at 77 K (Jo et al. 2019).

Table 2.2 Mechanical properties of high-entropy alloys at low temperatures (George et al. 2020)

Composition	*T*, K	σ, MPa	ε, %	Source
$Co_{20}Cr_{10}Fe_{45}Ni_{15}V_{10}$	77	1000	60	Jo et al. (2019)
$Co_{30}Cr_{10}Fe_{45}Ni_{5}V_{10}$	77	1500	50	Song et al. (2019)
$Co_{30}Cr_{10}Fe_{50}V_{10}$	77	2000	71	Kim et al. (2019)
$CoCrFeNiAl_{0.3}$	77	1355	66.5	Li et al. (2019)
CoCrFeNiMn	77	~1400	50	Sun et al. (2019)
CoCrFeNiMn	203	~1000	45	Sun et al. (2019)

σ—ultimate tensile strength (max.); ε—breaking deformation

The whole complex of properties makes HEAs highly perspective materials for use in conditions of Arctic.

The HEA's high-temperature mechanical properties (Rogachev 2020) are studied to a lesser extent. A high specific weight and brittleness of HEAs based on refractory metals are indicated as a main obstacle for their use in aerospace equipment whereas HEAs with a low specific weight possess no sufficient strength at high temperatures (Trofimenko et al. 2018). At the same time, the refractory HEAs possess good mechanical properties at room temperature: yield point is 930–1575 MPa, hardness is 335–705 HV, breaking deformation is ~30% (Yao et al. 2017). The compression tests of CoCrFeNiMn alloy carried out in temperature range 1023–1323 K at a deformation rate of 0.001–10 s^{-1} showed that the mechanism of deformation is the same as in pure metals with FCC-structure (Jeong et al. 2019).

References

Ahmad, A.S., Su, Y., Liu, S.Y., Ståhl, K., Wu, Y.D., Hui, X.D., Ruett, U., Gutowski, O., Glazyrin, K., Liermann, H.P., Franz, H., Wang, H., Wang, X.D., Cao, Q.P., Zhang, D.X., Jiang, J.Z.: J. Appl. Phys. **121**, 235901 (2017)

Bardziński, P.J., Błyskun, P.: Mater. Des. **103**, 377 (2016)

Bönisch, M., Wu, Y., Sehitoglu, H.: Acta Mater. **153**, 391 (2018)

Braeckman, B.R., Misják, F., Radnóczi, G., Caplovicová, M., Djemia, P., Tétard, F., Belliard, L., Depla, D.: Scr. Mater. **139**, 155 (2017)

Cai, Z., Cui, X., Jin, G., Liu, Z., Tan, N., Li, Y., Dong, M., Zhang, D.: Mater. Charact. **132**, 373 (2017)

Cakmak, G.: Adv. Mater. Sci. Eng. **2017**, 1950196 (2017)

Chang, Y.-J., Yeh, A.-C.: J. Alloys Compd. **653**, 379 (2015)

Chen, Z., Chen, W., Wu, B., Cao, X., Liu, L., Fu, Z.: Mater. Sci. Eng. A **648**, 217 (2015)

Chen, H., Kauffmann, A., Gorr, B., Schliephake, D., Seemüller, C., Wagner, J.N., Christ, H.-J., Heilmaier, M.: J. Alloys Compd. **661**, 206 (2016a)

Chen, J., Niu, P., Liu, Y., Lu, Y., Wang, X., Peng, Y., Liu, J.: Mater. Des. **94**, 39 (2016b)

Chen, X., Qi, J.Q., Sui, Y.W., He, Y.Z., Wei, F.X., Meng, Q.K., Sun, Z.: Mater. Sci. Eng. A **681**, 25 (2017)

Chen, R., Qin, G., Zheng, H., Wang, L., Su, Y., Chiu, Y.L., Ding, H., Guo, J., Fu, H.: Acta Mater. **144**, 129 (2018)

Choudhuri, D., Alam, T., Borkar, T., Gwalani, B., Mantri, A.S., Srinivasan, S.G., Gibson, M.A., Banerjee, R.: Scr. Mater. **100**, 36 (2015)

Daoud, H.M., Manzoni, A.M., Wanderka, N., Glatzel, U.: JOM **67**, 2271 (2015)

Dirras, G., Gubicza, J., Heczel, A., Lilensten, L., Couzinié, J.-P., Perrière, L., Guillot, I., Hocini, A.: Mater. Charact. **108**, 1 (2015)

Dirras, G., Lilensten, L., Djemia, P., Laurent-Brocq, M., Tingaud, D., Couzinié, J.-P., Perrière, L., Chauveau, T., Guillot, I.: Mater. Sci. Eng. A **654**, 30 (2016a)

Dirras, G., Couque, H., Lilensten, L., Heczel, A., Tingaud, D., Couzinié, J.-P., Perrière, L., Gubicza, J., Guillot, I.: Mater. Charact. **111**, 106 (2016b)

Feng, X.B., Zhang, J.Y., Wang, Y.Q., Hou, Z.Q., Wu, K., Liu, G., Sun, J.: Int. J. Plast. **95**, 264 (2017a)

Feng, X.B., Fu, W., Zhang, J.Y., Zhao, J.T., Li, J., Wu, K., Liu, G., Sun, J.: Scr. Mater. **139**, 71 (2017b)

Feng, X., Zhang, J., Xia, Z., Fu, W., Wu, K., Liu, G., Sun, J.: Mater. Lett. **210**, 84 (2018)

Fu, Z., Chen, W., Chen, Z., Wen, H., Lavernia, E.J.: Mater. Sci. Eng. A **619**, 137 (2014)

Fu, Z., Chen, W., Wen, H., Morgan, S., Chen, F., Zheng, B., Zhou, Y., Zhang, L., Lavernia, E.J.: Mater. Sci. Eng. A **644**, 10 (2015)

Fu, Z., Chen, W., Wen, H., Zhang, D., Chen, Z., Zheng, B., Zhou, Y., Lavernia, E.J.: Acta Mater. **107**, 59 (2016)

Fu, J.X., Cao, C.M., Tong, W., Hao, Y.X., Peng, L.M.: Mater. Sci. Eng. A **690**, 418 (2017)

Fu, Z., MacDonald, B.E., Zhang, D., Wu, B., Chen, W., Ivanisenko, J., Hahn, H., Lavernia, E.J.: Scr. Mater. **143**, 108 (2018)

Gali, A., George, E.P.: Intermetallics **39**, 74 (2013)

George, E.P., Curtin, W.A., Tasan, C.C.: Acta Mater. **188**, 435 (2020)

Gludovatz, B., Hohenwarter, A., Catoor, D., Chang, E.H., George, E.P., Ritchie, R.O.: Science **345**(80), 1153 (2014).

Gludovatz, B., George, E.P., Ritchie, R.O.: JOM **67**, 2262 (2015)

Gong, P., Jin, J., Deng, L., Wang, S., Gu, J., Yao, K., Wang, X.: Mater. Sci. Eng. A **688**, 174 (2017)

Guo, S.: Mater. Sci. Technol. **31**, 1223 (2015)

Guo, N.N., Wang, L., Luo, L.S., Li, X.Z., Su, Y.Q., Guo, J.J., Fu, H.Z.: Mater. Des. **81**, 87 (2015)

Guo, J., Huang, X., Huang, W.: J. Mater. Eng. Perform. **26**, 3071 (2017)

Gwalani, B., Soni, V., Lee, M., Mantri, S.A., Ren, Y., Banerjee, R.: Mater. Des. **121**, 254 (2017)

Gwalani, B., Gorsse, S., Choudhuri, D., Styles, M., Zheng, Y., Mishra, R.S., Banerjee, R.: Acta Mater. **153**, 169 (2018)

Haase, C., Barrales-Mora, L.A.: Acta Mater. **150**, 88 (2018)

Haglund, A., Koehler, M., Catoor, D., George, E.P., Keppens, V.: Intermetallics **58**, 62 (2015)

He, J.Y., Liu, W.H., Wang, H., Wu, Y., Liu, X.J., Nieh, T.G., Lu, Z.P.: Acta Mater. **62**, 105 (2014)

He, F., Wang, Z., Niu, S., Wu, Q., Li, J., Wang, J., Liu, C.T., Dang, Y.: J. Alloys Compd. **667**, 53 (2016a)

He, J.Y., Wang, H., Huang, H.L., Xu, X.D., Chen, M.W., Wu, Y., Liu, X.J., Nieh, T.G., An, K., Lu, Z.P.: Acta Mater. **102**, 187 (2016b)

He, J.Y., Wang, H., Wu, Y., Liu, X.J., Nieh, T.G., Lu, Z.P.: Mater. Sci. Eng. A **686**, 34 (2017a)

He, Q.F., Zeng, J.F., Wang, S., Ye, Y.F., Zhu, C., Nieh, T.G., Lu, Z.P., Yang, Y.: Mater. Res. Lett. **5**, 300 (2017b)

Hemphill, M.A., Yuan, T., Wang, G.Y., Yeh, J.W., Tsai, C.W., Chuang, A., Liaw, P.K.: Acta Mater. **60**, 5723 (2012)

Hong, S.I., Moon, J., Hong, S.K., Kim, H.S.: Mater. Sci. Eng. A **682**, 569 (2017)

Huo, W., Zhou, H., Fang, F., Xie, Z., Jiang, J.: Mater. Des. **134**, 226 (2017a)

Huo, W., Fang, F., Zhou, H., Xie, Z., Shang, J., Jiang, J.: Scr. Mater. **141**, 125 (2017b)

Ivchenko, M.V.: Structure, Phase Transformations and Properties of High-Entropy Equiatomic Metal Alloys Based on AlCrFeCoNiCu (2015)

Jang, M.J., Joo, S.-H., Tsai, C.-W., Yeh, J.-W., Kim, H.S.: Met. Mater. Int. **22**, 982 (2016)

Jeong, H.T., Park, H.K., Park, K., Na, T.W., Kim, W.J.: Mater. Sci. Eng. A **756**, 528 (2019)

Ji, X.: Int. J. Cast Met. Res. **28**, 229 (2015)
Jiang, H., Jiang, L., Han, K., Lu, Y., Wang, T., Cao, Z., Li, T.: J. Mater. Eng. Perform. **24**, 4594 (2015a)
Jiang, L., Cao, Z.Q., Jie, J.C., Zhang, J.J., Lu, Y.P., Wang, T.M., Li, T.J.: J. Alloys Compd. **649**, 585 (2015b)
Jiang, Z.J., He, J.Y., Wang, H.Y., Zhang, H.S., Lu, Z.P., Dai, L.H.: Mater. Res. Lett. **4**, 226 (2016a)
Jiang, H., Zhang, H., Huang, T., Lu, Y., Wang, T., Li, T.: Mater. Des. **109**, 539 (2016b)
Jiao, Z.-M., Ma, S.-G., Yuan, G.-Z., Wang, Z.-H., Yang, H.-J., Qiao, J.-W.: J. Mater. Eng. Perform. **24**, 3077 (2015a)
Jiao, Z.M., Chu, M.Y., Yang, H.J., Wang, Z.H., Qiao, J.W.: Mater. Sci. Technol. **31**, 1244 (2015b)
Jiao, Z.M., Ma, S.G., Chu, M.Y., Yang, H.J., Wang, Z.H., Zhang, Y., Qiao, J.W.: J. Mater. Eng. Perform. **25**, 451 (2016)
Jin, K., Guo, W., Lu, C., Ullah, M.W., Zhang, Y., Weber, W.J., Wang, L., Poplawsky, J.D., Bei, H.: Acta Mater. **121**, 365 (2016)
Jin, K., Mu, S., An, K., Porter, W.D., Samolyuk, G.D., Stocks, G.M., Bei, H.: Mater. Des. **117**, 185 (2017)
Jo, Y.H., Doh, K.-Y., Kim, D.G., Lee, K., Kim, D.W., Sung, H., Sohn, S.S., Lee, D., Kim, H.S., Lee, B.-J., Lee, S.: J. Alloys Compd. **809**, 151864 (2019)
Joo, S.-H., Kato, H., Jang, M.J., Moon, J., Tsai, C.W., Yeh, J.W., Kim, H.S.: Mater. Sci. Eng. A **689**, 122 (2017)
Joseph, J., Jarvis, T., Wu, X., Stanford, N., Hodgson, P., Fabijanic, D.M.: Mater. Sci. Eng. A **633**, 184 (2015)
Joseph, J., Stanford, N., Hodgson, P., Fabijanic, D.M.: Scr. Mater. **129**, 30 (2017a)
Joseph, J., Stanford, N., Hodgson, P., Fabijanic, D.M.: J. Alloys Compd. **726**, 885 (2017b)
Juan, C.-C., Tseng, K.-K., Hsu, W.-L., Tsai, M.-H., Tsai, C.-W., Lin, C.-M., Chen, S.-K., Lin, S.-J., Yeh, J.-W.: Mater. Lett. **175**, 284 (2016a)
Juan, C.-C., Tsai, M.-H., Tsai, C.-W., Hsu, W. L., Lin, C.-M., Chen, S.-K., Lin, S.-J., Yeh, J.-W.: Mater. Lett. **184**, 200 (2016b)
Kawasaki, M., Ahn, B., Kumar, P., Jang, J., Langdon, T.G.: Adv. Eng. Mater. **19**, 1600578 (2017)
Kilmametov, A., Kulagin, R., Mazilkin, A., Seils, S., Boll, T., Heilmaier, M., Hahn, H.: Scr. Mater. **158**, 29 (2019)
Kim, D.G., Jo, Y.H., Yang, J., Choi, W.-M., Kim, H.S., Lee, B.-J., Sohn, S.S., Lee, S.: Scr. Mater. **171**, 67 (2019)
Kireeva, I., Chumlyakov, Y., Pobedennaya, Z., Kuksgauzen, D., Karaman, I., Sehitoglu, H.: p. 020090 (2016)
Kireeva, I.V., Chumlyakov, Y.I., Pobedennaya, Z.V., Vyrodova, A.V., Karaman, I.: Mater. Sci. Eng. A **713**, 253 (2018)
Komarasamy, M., Kumar, N., Tang, Z., Mishra, R.S., Liaw, P.K.: Mater. Res. Lett. **3**, 30 (2015)
Koshelev, P.F.: Strength Mater. **3**, 286 (1971)
Kourov, N.I., Pushin, V.G., Korolev, A.V., Knyazev, Y.V., Ivchenko, M.V., Ustyugov, Y.M.: J. Alloys Compd. **636**, 304 (2015)
Kumar, N., Ying, Q., Nie, X., Mishra, R.S., Tang, Z., Liaw, P.K., Brennan, R.E., Doherty, K.J., Cho, K.C.: Mater. Des. **86**, 598 (2015)
Kuznetsov, A.V., Salishchev, G.A., Senkov, O.N., Stepanov, N.D., Shaisultanov, D.G.: Sci. Rec. Belgorod State Univ. Ser. Math. Phys. 182 (2012).
Kuznetsov, A.V., Shaysultanov, D.G., Stepanov, N.D., Salishchev, G.A., Senkov, O.N.: Mater. Sci. Eng. A **533**, 107 (2012b)
Laktionova, M.A., Tabachnikova, E.D., Tang, Z., Liaw, P.K.: Low Temp. Phys. **39**, 814 (2013)
Laplanche, G., Gadaud, P., Horst, O., Otto, F., Eggeler, G., George, E.P.: J. Alloys Compd. **623**, 348 (2015)
Laplanche, G., Kostka, A., Horst, O.M., Eggeler, G., George, E.P.: Acta Mater. **118**, 152 (2016)
Laplanche, G., Bonneville, J., Varvenne, C., Curtin, W.A., George, E.P.: Acta Mater. **143**, 257 (2018a)

Laplanche, G., Gadaud, P., Bärsch, C., Demtröder, K., Reinhart, C., Schreuer, J., George, E.P.: J. Alloys Compd. **746**, 244 (2018b)
Lee, C.-F., Shun, T.-T.: Mater. Charact. **114**, 179 (2016)
Lee, D.-H., Choi, I.-C., Seok, M.-Y., He, J., Lu, Z., Suh, J.-Y., Kawasaki, M., Langdon, T.G., Jang, J.: J. Mater. Res. **30**, 2804 (2015)
Lee, D.-H., Seok, M.-Y., Zhao, Y., Choi, I.-C., He, J., Lu, Z., Suh, J.-Y., Ramamurty, U., Kawasaki, M., Langdon, T.G., Jang, J.: Acta Mater. **109**, 314 (2016)
Lee, K.S., Kang, J.-H., Lim, K.R., Na, Y.S.: Mater. Charact. **132**, 162 (2017)
Lee, C., Song, G., Gao, M.C., Feng, R., Chen, P., Brechtl, J., Chen, Y., An, K., Guo, W., Poplawsky, J.D., Li, S., Samaei, A.T., Chen, W., Hu, A., Choo, H., Liaw, P.K.: Acta Mater. **160**, 158 (2018)
Lei, Z., Liu, X., Wu, Y., Wang, H., Jiang, S., Wang, S., Hui, X., Wu, Y., Gault, B., Kontis, P., Raabe, D., Gu, L., Zhang, Q., Chen, H., Wang, H., Liu, J., An, K., Zeng, Q., Nieh, T.-G., Lu, Z.: Nature **563**, 546 (2018)
Li, C., Xue, Y., Hua, M., Cao, T., Ma, L., Wang, L.: Mater. Des. **90**, 601 (2016a)
Li, J., Fang, Q., Liu, B., Liu, Y., Liu, Y.: RSC Adv. **6**, 76409 (2016b)
Li, X.C., Dou, D., Zheng, Z.Y., Li, J.C.: J. Mater. Eng. Perform. **25**, 2164 (2016c)
Li, J., Jia, W., Wang, J., Kou, H., Zhang, D., Beaugnon, E.: Mater. Des. **95**, 183 (2016d)
Li, P., Wang, A., Liu, C.T.: J. Alloys Compd. **694**, 55 (2017a)
Li, R., Wang, M., Yuan, T., Song, B., Shi, Y.: Metall. Mater. Trans. A **48**, 841 (2017b)
Li, Z., Tasan, C.C., Pradeep, K.G., Raabe, D.: Acta Mater. **131**, 323 (2017c)
Li, Z., Tasan, C.C., Springer, H., Gault, B., Raabe, D.: Sci. Rep. **7**, 40704 (2017d)
Li, Z., Zhao, S., Alotaibi, S.M., Liu, Y., Wang, B., Meyers, M.A.: Acta Mater. **151**, 424 (2018)
Li, Q., Zhang, T.W., Qiao, J.W., Ma, S.G., Zhao, D., Lu, P., Xu, B., Wang, Z.H.: Mater. Sci. Eng. A **767**, 138424 (2019)
Liang, Z.Y., De Hosson, J.T.M., Huang, M.X.: Acta Mater. **129**, 1 (2017)
Lim, K.R., Lee, K.S., Lee, J.S., Kim, J.Y., Chang, H.J., Na, Y.S.: J. Alloys Compd. **728**, 1235 (2017)
Lin, C.-W., Tsai, M.-H., Tsai, C.-W., Yeh, J.-W., Chen, S.-K.: Mater. Sci. Technol. **31**, 1165 (2015)
Liu, D., Cheng, J.B., Ling, H.: Mater. Sci. Technol. **31**, 1159 (2015a)
Liu, S., Gao, M.C., Liaw, P.K., Zhang, Y.: J. Alloys Compd. **619**, 610 (2015b)
Liu, X., Zhang, L., Xu, Y.: Appl. Phys. A **123**, 567 (2017a)
Liu, J., Chen, C., Xu, Y., Wu, S., Wang, G., Wang, H., Fang, Y., Meng, L.: Scr. Mater. **137**, 9 (2017b)
Liu, T.K., Wu, Z., Stoica, A.D., Xie, Q., Wu, W., Gao, Y.F., Bei, H., An, K.: Mater. Des. **131**, 419 (2017c)
Luo, H., Li, Z., Raabe, D.: Sci. Rep. **7**, 9892 (2017)
Ma, S.G., Zhang, S.F., Qiao, J.W., Wang, Z.H., Gao, M.C., Jiao, Z.M., Yang, H.J., Zhang, Y.: Intermetallics **54**, 104 (2014)
Ma, S.G., Qiao, J.W., Wang, Z.H., Yang, H.J., Zhang, Y.: Mater. Des. **88**, 1057 (2015)
Ma, S.G., Jiao, Z.M., Qiao, J.W., Yang, H.J., Zhang, Y., Wang, Z.H.: Mater. Sci. Eng. A **649**, 35 (2016)
Ma, L., Li, C., Jiang, Y., Zhou, J., Wang, L., Wang, F., Cao, T., Xue, Y.: J. Alloys Compd. **694**, 61 (2017)
Ma, Y., Wang, Q., Jiang, B.B., Li, C.L., Hao, J.M., Li, X.N., Dong, C., Nieh, T.G.: Acta Mater. **147**, 213 (2018)
Maier-Kiener, V., Schuh, B., George, E.P., Clemens, H., Hohenwarter, A.: Mater. Des. **115**, 479 (2017)
Maiti, S., Steurer, W.: Acta Mater. **106**, 87 (2016)
Meng, G., Yue, T.M., Lin, X., Yang, H., Xie, H., Ding, X.: Opt. Laser Technol. **70**, 119 (2015)
Meyers, M.A., Vöhringer, O., Lubarda, V.A.: Acta Mater. **49**, 4025 (2001)
Mills, W.J.: Int. Mater. Rev. **42**, 45 (1997)
Ming, K., Bi, X., Wang, J.: Scr. Mater. **137**, 88 (2017)
Moravcik, I., Cizek, J., Gavendova, P., Sheikh, S., Guo, S., Dlouhy, I.: Mater. Lett. **174**, 53 (2016)
Moravcik, I., Cizek, J., Zapletal, J., Kovacova, Z., Vesely, J., Minarik, P., Kitzmantel, M., Neubauer, E., Dlouhy, I.: Mater. Des. **119**, 141 (2017)

Mridha, S., Das, S., Aouadi, S., Mukherjee, S., Mishra, R.S.: JOM **67**, 2296 (2015)
Munitz, A., Salhov, S., Hayun, S., Frage, N.: J. Alloys Compd. **683**, 221 (2016)
Murty, B.S., Yeh, J.W., Ranganathan, S., Bhattacharjee, P.P.: High-Entropy Alloys, 2nd edn. Elsevier, Amsterdam (2019)
Ng, C., Guo, S., Luan, J., Wang, Q., Lu, J., Shi, S., Liu, C.T.: J. Alloys Compd. **584**, 530 (2014)
Niendorf, T., Wegener, T., Li, Z., Raabe, D.: Scr. Mater. **143**, 63 (2018)
Niu, S., Kou, H., Guo, T., Zhang, Y., Wang, J., Li, J.: Mater. Sci. Eng. A **671**, 82 (2016)
Niu, P.D., Li, R.D., Yuan, T.C., Zhu, S.Y., Chen, C., Wang, M.B., Huang, L.: Intermetallics **104**, 24 (2019)
Ogura, M., Fukushima, T., Zeller, R., Dederichs, P.H.: J. Alloys Compd. **715**, 454 (2017)
Okamoto, N.L., Fujimoto, S., Kambara, Y., Kawamura, M., Chen, Z.M.T., Matsunoshita, H., Tanaka, K., Inui, H., George, E.P.: Sci. Rep. **6**, 35863 (2016)
Osintsev, K.A., Gromov, V.E., Konovalov, S.V., Ivanov, Y.F., Panchenko, I.A.: Izv. Ferr. Metall. 1 (2021)
Otto, F., Dlouhý, A., Somsen, C., Bei, H., Eggeler, G., George, E.P.: Acta Mater. **61**, 5743 (2013)
Park, N., Li, X., Tsuji, N.: JOM **67**, 2303 (2015)
Pi, J., Wang, Z., He, X., Bai, Y.: J. Mater. Eng. Perform. **25**, 76 (2016)
Pickering, E.J., Muñoz-Moreno, R., Stone, H.J., Jones, N.G.: Scr. Mater. **113**, 106 (2016)
Pogrebnjak, A.D., Bagdasaryan, A.A., Yakushchenko, I.V., Beresnev, V.M.: Russ. Chem. Rev. **83**, 1027 (2014)
Pogrebnyak, A.D., Bagdasapyan, A.A., Yakushchenko, I.V., Beresnev, V.M.: Russ. Chem. Rev. **83**, 1027 (2014)
Prasad, H., Singh, S., Panigrahi, B.B.: J. Alloys Compd. **692**, 720 (2017)
Qi L., Chrzan, D.C.: Phys. Rev. Lett. **112**, 115503 (2014)
Qiao, J.W., Ma, S.G., Huang, E.W., Chuang, C.P., Liaw, P.K., Zhang, Y.: Mater. Sci. Forum **688**, 419 (2011)
Raghavan, R., Kirchlechner, C., Jaya, B.N., Feuerbacher, M., Dehm, G.: Scr. Mater. **129**, 52 (2017)
Rao, Z., Wang, X., Wang, Q., Liu, T., Chen, X., Wang, L., Hui, X.: Adv. Eng. Mater. **19**, 1600726 (2017)
Reddy, S.R., Bapari, S., Bhattacharjee, P.P., Chokshi, A.H.: Mater. Res. Lett. **5**, 408 (2017)
Rogachev, A.S.: Phys. Met. Mater. Sci. **121**, 807 (2020)
Rogal, Ł, Morgiel, J., Świątek, Z., Czerwiński, F.: Mater. Sci. Eng. A **651**, 590 (2016)
Salishchev, G.A., Tikhonovsky, M.A., Shaysultanov, D.G., Stepanov, N.D., Kuznetsov, A.V., Kolodiy, I.V., Tortika, A.S., Senkov, O.N.: J. Alloys Compd. **591**, 11 (2014)
Sathiyamoorthi, P., Basu, J., Kashyap, S., Pradeep, K.G., Kottada, R.S.: Mater. Des. **134**, 426 (2017)
Schuh, B., Mendez-Martin, F., Völker, B., George, E.P., Clemens, H., Pippan, R., Hohenwarter, A.: Acta Mater. **96**, 258 (2015)
Schuh, B., Völker, B., Todt, J., Schell, N., Perrière, L., Li, J., Couzinié, J.P., Hohenwarter, A.: Acta Mater. **142**, 201 (2018)
Senkov, O.N., Semiatin, S.L.: J. Alloys Compd. **649**, 1110 (2015)
Senkov, O.N., Wilks, G.B., Miracle, D.B., Chuang, C.P., Liaw, P.K.: Intermetallics **18**, 1758 (2010)
Senkov, O.N., Wilks, G.B., Scott, J.M., Miracle, D.B.: Intermetallics **19**, 698 (2011a)
Senkov, O.N., Scott, J.M., Senkova, S.V., Miracle, D.B., Woodward, C.F.: J. Alloys Compd. **509**, 6043 (2011b)
Senkov, O., Scott, J., Senkova, S., Meisenkothen, F., Miracle, D., Woodward, C.: J. Mater. Sci. **47**, 4062 (2012)
Senkov, O.N., Senkova, S.V., Miracle, D.B., Woodward, C.: Mater. Sci. Eng. A **565**, 51 (2013)
Seol, J.B., Bae, J.W., Li, Z., Chan Han, J., Kim, J.G., Raabe, D., Kim, H.S.: Acta Mater. **151**, 366 (2018)
Shahmir, H., He, J., Lu, Z., Kawasaki, M., Langdon, T.G.: Mater. Sci. Eng. A **685**, 342 (2017)
Shaisultanov, D.G.: Structure and Mechanical Properties of CoCrFeNiX System (X = Mn, V, Mn and V, Al and Cu) High-Entropy Alloys (2015)

Shaysultanov, D.G., Stepanov, N.D., Salishchev, G.A., Tikhonovsky, M.A.: Phys. Met. Metallogr. **118**, 579 (2017a)
Shaysultanov, D.G., Salishchev, G.A., Ivanisenko, Y.V., Zherebtsov, S.V., Tikhonovsky, M.A., Stepanov, N.D.: J. Alloys Compd. **705**, 756 (2017b)
Slone, C.E., Chakraborty, S., Miao, J., George, E.P., Mills, M.J., Niezgoda, S.R.: Acta Mater. **158**, 38 (2018)
Song, H., Yang, J., Jo, Y.H., Song, T., Kim, H.S., Lee, B.-J., Lee, S.: J. Alloys Compd. **797**, 465 (2019)
Sosa, J.M., Jensen, J.K., Huber, D.E., Viswanathan, G.B., Gibson, M.A., Fraser, H.L.: Mater. Sci. Technol. **31**, 1250 (2015)
Stepanov, N.D., Shaysultanov, D.G., Tikhonovsky, M.A., Salishchev, G.A.: Mater. Des. **87**, 60 (2015a)
Stepanov, N.D., Yurchenko, N.Y., Skibin, D.V., Tikhonovsky, M.A., Salishchev, G.A.: J. Alloys Compd. **652**, 266 (2015b)
Stepanov, N.D., Shaysultanov, D.G., Chernichenko, R.S., Yurchenko, N.Y., Zherebtsov, S.V., Tikhonovsky, M.A., Salishchev, G.A.: J. Alloys Compd. **693**, 394 (2017)
Strife, J., Passoja, D.: Metall. Trans. **11**, 1341 (1980)
Sun, S.J., Tian, Y.Z., Lin, H.R., Dong, X.G., Wang, Y.H., Zhang, Z.J., Zhang, Z.F.: Mater. Des. **133**, 122 (2017)
Sun, S.J., Tian, Y.Z., Lin, H.R., Dong, X.G., Wang, Y.H., Wang, Z.J., Zhang, Z.F.: J. Alloys Compd. **806**, 992 (2019)
Tabachnikova, E.D., Podolskiy, A.V., Laktionova, M.O., Bereznaia, N.A., Tikhonovsky, M.A., Tortika, A.S.: J. Alloys Compd. **698**, 501 (2017)
Tariq, N.H., Naeem, M., Hasan, B.A., Akhter, J.I., Siddique, M.: J. Alloys Compd. **556**, 79 (2013)
Tazuddin, A., Biswas, K., Gurao, N.P.: Mater. Sci. Eng. A **657**, 224 (2016)
Tian, L., Jiao, Z.M., Yuan, G.Z., Ma, S.G., Wang, Z.H., Yang, H.J., Zhang, Y., Qiao, J.W.: J. Mater. Eng. Perform. **25**, 2255 (2016)
Tian, L.-Y., Wang, G., Harris, J.S., Irving, D.L., Zhao, J., Vitos, L.: Mater. Des. **114**, 243 (2017)
Trofimenko, N.N., Efimochkin, I.Y., Bol'shakova, A.N.: Aviat. Mater. Technol. **51**, 3 (2018)
Tsao, T.-K., Yeh, A.-C., Murakami, H.: Metall. Mater. Trans. A **48**, 2435 (2017a)
Tsao, T.-K., Yeh, A.-C., Kuo, C.-M., Kakehi, K., Murakami, H., Yeh, J.-W., Jian, S.-R.: Sci. Rep. **7**, 12658 (2017b)
Tung, C.-C., Yeh, J.-W., Shun, T., Chen, S.-K., Huang, Y.-S., Chen, H.-C.: Mater. Lett. **61**, 1 (2007)
Varvenne, C., Curtin, W.A.: Scr. Mater. **138**, 92 (2017)
Varvenne, C., Curtin, W.A.: Scr. Mater. **142**, 92 (2018)
Varvenne, C., Luque, A., Curtin, W.A.: Acta Mater. **119**, 242 (2016)
Vida, Á., Chinh, N.Q., Lendvai, J., Heczel, A., Varga, L.K.: Mater. Lett. **195**, 14 (2017)
Wang, S.-P., Xu, J.: Mater. Sci. Eng. C **73**, 80 (2017)
Wang, X.F., Zhang, Y., Qiao, Y., Chen, G.L.: Intermetallics **15**, 357 (2007)
Wang, W.-R., Wang, W.-L., Yeh, J.-W.: J. Alloys Compd. **589**, 143 (2014)
Wang, X.-R., He, P., Lin, T.-S., Wang, Z.-Q.: Mater. Sci. Technol. **31**, 1842 (2015)
Wang, B., Fu, A., Huang, X., Liu, B., Liu, Y., Li, Z., Zan, X.: J. Mater. Eng. Perform. **25**, 2985 (2016a)
Wang, Z., Baker, I., Cai, Z., Chen, S., Poplawsky, J.D., Guo, W.: Acta Mater. **120**, 228 (2016b)
Wang, X.-R., Wang, Z.-Q., Lin, T.-S., He, P.: Mater. Sci. Technol. **32**, 1289 (2016c)
Wang, X.-R., Wang, Z.-Q., Lin, T.-S., He, P., Sekulic, D.P.: J. Mater. Eng. Perform. **25**, 2053 (2016d)
Wang, R., Zhang, K., Davies, C., Wu, X.: J. Alloys Compd. **694**, 971 (2017a)
Wang, J., Liu, Y., Liu, B., Wang, Y., Cao, Y., Li, T., Zhou, R.: Mater. Sci. Eng. A **689**, 233 (2017b)
Wang, Y., Liu, B., Yan, K., Wang, M., Kabra, S., Chiu, Y.-L., Dye, D., Lee, P.D., Liu, Y., Cai, B.: Acta Mater. **154**, 79 (2018)
Wani, I.S., Bhattacharjee, T., Sheikh, S., Bhattacharjee, P.P., Guo, S., Tsuji, N.: Mater. Sci. Eng. A **675**, 99 (2016)
Wei, S., He, F., Tasan, C.C.: J. Mater. Res. **33**, 2924 (2018)

Wu, Z., Bei, H., Pharr, G.M., George, E.P.: Acta Mater. **81**, 428 (2014)
Wu, S.W., Wang, G., Yi, J., Jia, Y.D., Hussain, I., Zhai, Q.J., Liaw, P.K.: Mater. Res. Lett. **5**, 276 (2017a)
Wu, Z., Troparevsky, M.C., Gao, Y.F., Morris, J.R., Stocks, G.M., Bei, H.: Curr. Opin. Solid State Mater. Sci. **21**, 267 (2017b)
Xian, X., Zhong, Z., Zhang, B., Song, K., Chen, C., Wang, S., Cheng, J., Wu, Y.: Mater. Des. **121**, 229 (2017)
Xiao, D.H.H., Zhou, P.F.F., Wu, W.Q.Q., Diao, H.Y.Y., Gao, M.C.C., Song, M., Liaw, P.K.K.: Mater. Des. **116**, 438 (2017)
Xu, X.D., Liu, P., Tang, Z., Hirata, A., Song, S.X., Nieh, T.G., Liaw, P.K., Liu, C.T., Chen, M.W.: Acta Mater. **144**, 107 (2018)
Yang, X., Zhang, Y., Liaw, P.K.: Procedia Eng. **36**, 292 (2012)
Yang, T., Xia, S., Liu, S., Wang, C., Liu, S., Zhang, Y., Xue, J., Yan, S., Wang, Y.: Mater. Sci. Eng. A **648**, 15 (2015)
Yang, T., Tang, Z., Xie, X., Carroll, R., Wang, G., Wang, Y., Dahmen, K.A., Liaw, P.K., Zhang, Y.: Mater. Sci. Eng. A **684**, 552 (2017)
Yang, X., Zhou, Y., Xi, S., Chen, Z., Wei, P., He, C., Li, T., Gao, Y., Wu, H.: Mater. Sci. Eng. A **767**, 138394 (2019)
Yang, X., Zhou, Y., Xi, S., Chen, Z., Wei, P., He, C., Li, T., Gao, Y., Wu, H.: Mater. Sci. Eng. A **767**, 138382 (2019)
Yao, H.W., Qiao, J.W., Hawk, J.A., Zhou, H.F., Chen, M.W., Gao, M.C.: J. Alloys Compd. **696**, 1139 (2017)
Yao, H., Tan, Z., He, D., Zhou, Z., Zhou, Z., Xue, Y., Cui, L., Chen, L., Wang, G., Yang, Y.: J. Alloys Compd. **813**, 152196 (2020)
Yasuda, H.Y., Shigeno, K., Nagase, T.: Scr. Mater. **108**, 80 (2015)
Yeh, J.-W.: JOM **67**, 2254 (2015)
Yeh, J.W., Chen, Y.L., Lin, S.J., Chen, S.K.: Mater. Sci. Forum **560**, 1 (2007)
Yu, P.F., Cheng, H., Zhang, L.J., Zhang, H., Jing, Q., Ma, M.Z., Liaw, P.K., Li, G., Liu, R.P.: Mater. Sci. Eng. A **655**, 283 (2016)
Yurchenko, N.Y., Stepanov, N.D., Shaysultanov, D.G., Tikhonovsky, M.A., Salishchev, G.A.: Mater. Charact. **121**, 125 (2016)
Yurchenko, N.Y., Stepanov, N.D., Zherebtsov, S.V., Tikhonovsky, M.A., Salishchev, G.A.: Mater. Sci. Eng. A **704**, 82 (2017)
Yurkova, A.I., Chernyavskii, V.V., Gorban V.F.: Powder Metall. Met. Ceram. **55**, 152 (2016).
Zaddach, A.J., Niu, C., Koch, C.C., Irving, D.L.: JOM **65**, 1780 (2013)
Zhang, Y.: High-Entropy Materials. Springer Singapore, Singapore (2019)
Zhang, H.F.Z.W.Q., Lou, C.S., Wu, X.C., Fu, H.M.: Mater. Sci. Technol. **31**, 1342 (2015)
Zhang, H., Siu, K.W., Liao, W., Wang, Q., Yang, Y., Lu, Y.: Mater. Res. Express **3**, 094002 (2016)
Zhang, Y., Liu, Y., Li, Y., Chen, X., Zhang, H.: Mater. Lett. **174**, 82 (2016b)
Zhang, L., Yu, P., Cheng, H., Zhang, H., Diao, H., Shi, Y., Chen, B., Chen, P., Feng, R., Bai, J., Jing, Q., Ma, M., Liaw, P.K., Li, G., Liu, R.: Metall. Mater. Trans. A **47**, 5871 (2016c)
Zhang, M., Zhou, X., Li, J.: J. Mater. Eng. Perform. **26**, 3657 (2017a)
Zhang, H., Tang, H., He, Y.Z., Zhang, J.L., Li, W.H., Guo, S.: JOM **69**, 2078 (2017b)
Zhang, Z., Zhang, H., Tang, Y., Zhu, L., Ye, Y., Li, S., Bai, S.: Mater. Des. **133**, 435 (2017c)
Zhang, Y., Wang, S., Jiang, S., Zhu, X., Sun, D.: J. Mater. Eng. Perform. **26**, 41 (2017d)
Zhou, Y.J., Zhang, Y., Wang, Y.L., Chen, G.L.: Appl. Phys. Lett. **90**, 181904 (2007)
Zhu, Z.G., Ma, K.H., Yang, X., Shek, C.H.: J. Alloys Compd. **695**, 2945 (2017)

Chapter 3
HEAs' Stability

Abstract In this chapter, the problem of stability or metastability of HEAs was analysed. This problem is one of the main ones in analysing the influence of temperature on a whole range of properties. The factors determining the structure and stability of multicomponent disordered solid solutions are analysed. The concept of temperature domains of the existence of stability and metastability of phases is considered. It is noted that HEAs based on refractory metals are metastable. The Research is supported by the Russian Science Foundation (project No. 20-19-00452).

The problem of stability or metastability of HEAs is one of the main one in analysing a temperature effect on a whole complex of properties (Rogachev 2020). On HEA's crystallisation the atoms from the melting pack into simple crystal structures: FCC, BCC, or HCP (Cantor 2014). Atoms of different types locate randomly in crystal lattice sites, i.e. HEA is a disordered substitutional solid solution. A disordered position of all atoms in crystal lattice sites results in increased configurational entropy of the phase that gave the name to the class of materials (Rogachev 2020).

One of the most effective tools to determine a phase stability of the state is a phase diagram. To assess phase stability in the $Al_{0.5}CoCrCuFeNi$ high-entropy alloy being formed as a result of the thermal effect the authors of the paper (Ng et al. 2012) performed thermodynamic calculations. It was supposed that at the corresponding temperature the phases obtained as a result of annealing were equilibrium ones. Prerequisites for such an assumption were the following facts: firstly, a presence of equilibrium phases in definite temperature conditions (that is consistent with experimental results); secondly, the annealing processes of HEAs are equivalent to quenching processes of conventional alloys. Due to a weak diffusion in HEA a cooling proceeds quickly, which prevents the termination of phase transformation with the formation of new phases (Pogrebnyak et al. 2014).

The majority of papers are concerned with research on the stability of HEAs belonging to the family of transition 3d-metals CoCrFeNiX, where x = Al, Ti, Cu, V or Mn, whose base is a disordered substitutional solid solution with FCC-structure CoCrFeNi (Rogachev 2020). But the question was not cleared up whether two FCC-phases had formed as a result of long-term ageing of single-phase metastable alloy

V. E. Gromov et al., *Structure and Properties of High-Entropy Alloys*, Advanced Structured Materials 107, https://doi.org/10.1007/978-3-030-78364-8_3

(supersaturated solid solution) or they melt already on crystallisation of the melt and remain stable later on. An intermediate variant is possible on crystallisation two phases are formed they are close in composition and lattice parameter but in the process of long-term annealing the differences grow between them (Rogachev 2020).

In the author's opinion of the review (Rogachev 2020) the addition of the fifth element to a nucleus of four elements results, as a rule, to the increase in distortions of crystal structure because the sizes of the fifth atom are different from a very close to each other atom sizes of Co, Cr, Fe and Ni. It should decrease in HEA stability in the transition from four-component to five-component systems, while the entropy factor, on the contrary, stabilizes the phases when the amount of components is increased (Rogachev 2020).

In Cantor classical alloy CoCrFeNiMn the Mn diameter (0.274 nm) is larger than that of four other elements (0.248–0.250 nm). The results of the researches (Pickering et al. 2016; He et al. 2014; Otto et al. 2016; Laplanche et al. 2018; Zaddach et al. 2015) testify to the inconsistency of the statements about a stability of this alloy.

It was noted that secondary phases form very slowly in coarse-grained alloys and much faster in fine-grained alloys subjected to severe plastic deformation. It may be connected with the increase in the specific area of grain boundaries along which diffusion proceeds or with the increase in internal energy due to severe plastic deformation that creates additional motive force of σ-phases precipitation. It can also be noted that a multicomponent solid solution decomposes only at relatively low ageing (annealing) temperatures, therefore, it is thermodynamically stable at a higher temperature (Rogachev 2020).

There appeared a rather large number of researches on HEA stability in the Co–Cr–Fe–Ni–Mn system in relation of components being different from equiatomic one. As precipitation of secondary phases based on chromium results in the decrease in the content of this metal in the HES matrix, a number of works are devoted to the study of compositions in which the total chromium content is increased at the expense of nickel, for example, $Co_{20}Cr_{26}Fe_{20}Ni_{14}Mn_{20}$ (Laplanche et al. 2018; Zaddach et al. 2015).

A precipitation of inclusions in CoCrFeNiMn alloy proceeds much faster if it was subjected to severe plastic deformation. According to the data of small-angle scattering of synchrotron radiation the inclusions with ~1.2 nm radius, formed after cold rolling of cast alloy, grow quickly during the first 30 min of annealing at 773 K and reach a saturation in 60 min (Huang et al. 2019).

Summing up the data on Cantor HEA stability the author (Rogachev 2020) (George et al. 2020) states that: (a) annealing of the alloy at a temperature of 1073 K and below results in precipitation of inclusions enriched in Cr and Mn in the alloy as well as intermetallic σ-phases; (b) the excess of Cr and Mn in the alloy, as well as its severe plastic deformation accelerate a formation of secondary phases; (c) the decrease in Mn content and increase in Co content stabilize FCC-phase; (d) if annealing occurs at temperature of 1173 K the precipitation of secondary phases fail to proceed during 500 days; (e) despite the formation of secondary phase inclusions, the main matrix phase remains a five-component solid solution with FCC-structure; (f) there is a good probability of the existence of two high-entropy FCC phases whose lattice constants differ by 0.2–1.1%, the phase with a more compact packing being more stable.

In the Co–Cr–Fe–Ni–Al system, a comparatively rapid structural transformations and precipitation of secondary phases are observed both in cast alloys and in those obtained by mechanical melting (Rogachev 2020).

Stability of the alloys based on refractory metals is yet not investigated in sufficient detail (Tang et al. 2019; Wang et al. 2019a, b; Pacheco et al. 2019; Poulia et al. 2018; Yao et al. 2018; Wu et al. 2018), but the experimental and theoretical data being available favour a metastability of the alloys. In the review (Rogachev 2020), it is noted that along with the investigation into the stability of HEA main families there is a quite large number of recently published papers wherein a stability of heterogeneous alloys was experimentally studied including those uniting the metals from different families. Table 3.1 presents only the main facts concerning the phase transformation in annealing.

Today four main approaches to the solution of fundamental problems of HEA stability (Rogachev 2020) may be distinguished. The first approach may be termed as semi-empirical. It consists of searching for regularities among a large number of experimental data and formulation of stability criteria. The second approach to solve the problem of HEA stability is based on quantum–mechanical calculations of crystal and electron structure of alloys proceeding from so-called first principles (laws of conservation, Schrödinger equation).

The third approach includes the methods of computer modelling (methods of molecular dynamics, Monte Carlo, etc.) which still begin to develop. However, large computer capacities are required for modelling of systems from millions of atoms therefore real simulation of processes in HEA is possible so far for short intervals of time (Rogachev 2020).

The fourth from the approaches being considered might be called the first in importance because it is thermodynamic calculations. Whatever criteria and methods are used, as the final result, a stability of any system means the absolute minimum of free energy. The program CALPHAD (CALculation of PHAse Diagrams) based on the principle is still more frequently used in HEA research (Rogachev 2020).

Table 3.1 Thermal stability of phase composition of some HEAs

Alloy	Methods	Initial phases	Annealing conditions	Phases after annealing	References
Five-component alloys					
$CoCrFeNiNb_{0.15–0.45}$	AM	FCC + Laves phases	773 and 1073 K, 12 h	Increasing Laves phases	Zhang et al. (2018)
$CoCrFeNiNb_{0.45}$	IM	FCC + C14 Laves[a]	873–1373 K, 4 h	FCC + C14: stable	Jiang et al. (2019)
$CrFeMnTiAl_{1.5}$	AM	BCC + $L2_1$ + C14	1073–1473 K, 168 h; 1273 K. 504 h	Changes quantity of $L2_1$ Depends on *T*	Feng et al. (2018)
CoFeNiMnAl	AM	*B*2	1323 K, 50 h	*B*2 + FCC	Karati et al. (2019)
CoFeNiMnCu	AM	Two FCC phases	1273 K, 24 h	**Single FCC**	(Shim et al. 2019)
AlCoCuNiZn	MM	FCC(β)	673–1273 K, 48 h	FCC(γ) + $L1_2$(α)	Mohanty et al. (2017)
$AlNbTiVCr_x$	AM	*B*2(+Cr_2Nb C14, at $x \geq 1$)	1073–1273 K, 100 h	Nb_2Al σ-phase, increasing in fraction of C14	Yurchenko et al. (2018)
$AlNbTiVZr_x$	AM	*B*2 + Zr_5Al_3 + C14	1073–1273 K, 100 h	**Almost stable**	Yurchenko et al. (2018)
$(CoFeNiMo)_{90}Al_{10}$	AM	$L1_2$ + μ(*Rm*3*m*)	1173 K up to 100 h	Nano-CoMo (?)	Chen et al. (2019)
$(CoFeNiMo)_{90}Cr_{10}$	AM	FCC + μ + σ-phases	1173 K p to 100 h	Layers, precipitations	Chen et al. (2019)

Methods Methods of preparation: *AM* Arc melting; *IM* Induction melting; *MM* Mechanical melting
[a]Apparently, Laves phase has a composition of Fe_2Nb in these alloys

References

Cantor, B.: Entropy **16**, 4749 (2014)
Chen, C., Liu, N., Zhang, J., Cao, J., Wang, L., Xiang, H.: Mater. Sci. Technol. **35**, 1883 (2019)
Feng, R., Gao, M.C., Zhang, C., Guo, W., Poplawsky, J.D., Zhang, F., Hawk, J.A., Neuefeind, J.C., Ren, Y., Liaw, P.K.: Acta Mater. **146**, 280 (2018)
George, E.P., Curtin, W.A., Tasan, C.C.: Acta Mater. **188**, 435 (2020)
He, J.Y., Zhu, C., Zhou, D.Q., Liu, W.H., Nieh, T.G., Lu, Z.P.: Intermetallics **55**, 9 (2014)
Huang, Y.-C., Tsao, C.-S., Wu, S.-K., Lin, C., Chen, C.-H.: Intermetallics **105**, 146 (2019)
Jiang, H., Qiao, D., Lu, Y., Ren, Z., Cao, Z., Wang, T., Li, T.: Scr. Mater. **165**, 145 (2019)
Karati, A., Guruvidyathri, K., Hariharan, V.S., Murty, B.S.: Scr. Mater. **162**, 465 (2019)
Laplanche, G., Berglund, S., Reinhart, C., Kostka, A., Fox, F., George, E.P.: Acta Mater. **161**, 338 (2018)
Mohanty, S., Gurao, N.P., Padaikathan, P., Biswas, K.: Mater. Charact. **129**, 127 (2017)

Ng, C., Guo, S., Luan, J., Shi, S., Liu, C.: Intermetallics **31**, 165 (2012)

Otto, F., Dlouhý, A., Pradeep, K.G., Kuběnová, M., Raabe, D., Eggeler, G., George, E.P.: Acta Mater. **112**, 40 (2016)

Pacheco, U.J.V., Lindwall, G., Karlsson, D., Cedervall, J., Fritze, S., Ek, G., Berastegui, P., Sahlberg, M.: Inorg. Chem. **58**, 811 (2019)

Pickering, E.J., Muñoz-Moreno, R., Stone, H.J., Jones, N.G.: Scr. Mater. **113**, 106 (2016)

Pogrebnyak, A.D., Bagdasapyan, A.A., Yakushchenko, I.V., Beresnev, V.M.: Russ. Chem. Rev. **83**, 1027 (2014)

Poulia, A., Georgatis, E., Mathiou, C., Karantzalis, A.E.: Mater. Chem. Phys. **210**, 251 (2018)

Rogachev, A.S.: Phys. Met. Mater. Sci. **121**, 807 (2020)

Shim, S.H., Oh, S.M., Lee, J., Hong, S.-K., Hong, S.I.: Mater. Sci. Eng. A **762**, 138120 (2019)

Tang, Y, Wang, R., Li, S., Liu, X., Ye, Y., Zhu, L., Bai, S., Xiao, B.: Mater. Des. **181**, 107928 (2019)

Wang, R., Tang, Y., Li, S., Zhang, H., Ye, Y., Zhu, L., Ai, Y., Bai, S.: Mater. Des. **162**, 256 (2019a)

Wang, G., Liu, Q., Yang, J., Li, X., Sui, X., Gu, Y., Liu, Y.: Int. J. Refract. Met. Hard Mater. **84**, 104988 (2019b)

Wu, Y., Si, J., Lin, D., Wang, T., Wang, W.Y., Wang, Y., Liu, Z., Hui, X.: Mater. Sci. Eng. A **724**, 249 (2018)

Yao, J.Q., Liu, X.W., Gao, N., Jiang, Q.H., Li, N., Liu, G., Zhang, W.B., Fan, Z.T.: Intermetallics **98**, 79 (2018)

Yurchenko, N.Y., Stepanov, N.D., Gridneva, A.O., Mishunin, M.V., Salishchev, G.A., Zherebtsov, S.V.: J. Alloys Compd. **757**, 403 (2018)

Zaddach, A.J., Scattergood, R.O., Koch, C.C.: Mater. Sci. Eng. A **636**, 373 (2015)

Zhang, M., Zhang, L., Liaw, P., Li, G., Liu, R.: J. Mater. Res. **33**, 1 (2018)

Chapter 4
Prospects of High-Entropy Alloys Application

Abstract In this chapter, the prospects of high-entropy alloys' application were observed. There are several the most promising spheres of applications of HEAs such as nuclear power industry, as they have high resistance to radiation, as refractory (due to high strength at elevated temperatures and resistance to oxidation), soft magnetic materials (exhibit superparamagnetic, ferromagnetic properties), as well as materials used in the tool industry (have a low coefficient of friction, wear resistance, an optimal ratio of strength and ductility), corrosion-resistant materials, in the aerospace industry (low density) and as materials for storing hydrogen. The Research is supported by the Russian Scientific Foundation (project No. 20-19-00452).

Depending on composition and microstructure the high-entropy alloys may possess a combination of attracting properties such as high strength, corrosion resistance, wear resistance, plasticity, heat resistance, high hardness (Shaisultanov 2015).

The majority of papers having been analysed in previous sections contain the results of investigations into HEA unique properties but also the examples of their practical application. The production of cast HEAs for commercial purposes has already begun, but so far it is premature to judge the purposeful full-scale introduction of HEAs into the industry. At the given stage of the scientific development of HEA, it is more correct to favour the nearest prospects for application. The high operational properties of high-entropy alloys make them potentially useful for application as tools, press moulds, stamps, mechanical parts and furnace members which require high strength, heat resistance, resistance to oxidation and wear. These also possess excellent corrosion resistance and may be used as anticorrosion high-strength materials at chemical plants, foundries and even as pipelines and pump members being operated in sea water. Besides, the development of technologies of coating will contribute to further extension of HEA'S use as functional coatings will contribute to such as diffusion barriers for copper compounds in superlarge integrated circuits and as soft magnetosoft films for high-frequency communication (Shaisultanov 2015).

We shall present some examples. For the possibility of use of heat resistant HEAs as heat barriers for supersonic planes, blades for gas turbines, etc., it is necessary to follow the requirements: working temperature is from 20 to 1500 °C; yield point is above 400 MPa; elongation >6% is a good resistance to oxidation. The investigation

V. E. Gromov et al., *Structure and Properties of High-Entropy Alloys*, Advanced Structured Materials 107, https://doi.org/10.1007/978-3-030-78364-8_4

into mechanical properties of high-entropy alloys (NbMoWTa and VNbMoWTa) showed that in temperature interval T = 800 – 1600 °C, the HEA yield point is much higher than that in superalloys (Shaisultanov 2015).

Investigation into $Co_{1.5}CrFeNi_{1.5}Ti_{0.5}$ alloy showed that in cast state it has a hardness of 378 HV, and after ageing at 800 °C for 5 h it is possible to obtain a maximum hardness being equal 513 HV (Murtu et al. 2014). In addition, the alloy exhibits less magnetisation and high specific resistance after ageing. The alloy has negligible losses on eddy currents in an alternating magnetic field. It also shows a better resistance to abrasion wear and a higher corrosion resistance in $0.5H_2SO_4 + 10.5$ NaCl solution. It may be used under severe conditions as members of underground electric pumps used in a system of oil well. Due to its high-temperature strength and resistance to oxidation the alloy may be used as spare parts such as connecting rods and high-temperature tensile holders in tension testing machines to 1000 °C at which its high-temperature hardness remains high – about 250 HV that is higher than in the majority of commercial heat resistant alloys.

In a review (Pogrebnyak et al. 2014), it is noted that high-entropy alloys may be used in nuclear power engineering, in particular, because these possess a high resistance to radiation (Egami et al. 2014), as refractory (due to high strength at elevated temperature and resistance to oxidation), magnetosoft materials (exhibit superparamagnetic, ferromagnetic properties) as well as the materials used in tool industry (have a low coefficient of friction, resistant to wear, optimal ratio of strength and plasticity), corrosion-resistant materials, in aerospace industry (low density) and as materials for hydrogen storage (Meyers et al. 2001).

One of the perspective fields of application of nitride coatings based on high-entropy alloys is biomedicine. Protective coatings for biomedical application should have a low modulus of elasticity, high chemical stability, wear and corrosion resistance in physiological media, low friction coefficient, biological compatibility and excellent adhesion to surface on which protective coatings are precipitated (Pogrebnyak et al. 2014).

The author of the review (Rogachev 2020) considers that the most perspective direction of HEA application may become magneto soft HEAs (Zuo et al. 2017; Mishra and Shahi 2020; Shkodich et al. 2020) and alloys for operation at low temperatures (Zhang 2019). There is information on HEAs of superconductors such as ReNbTiZrHf, LaCePrNdSmFeBeS (Vrtnik et al. 2017; Marik et al. 2018, 2019; Sogabe et al. 2018). It is clear that HEA application will not be limited by the fields and will be much wider as the new HEAs and their properties are studied.

4.1 Conclusion

The history of the development of metallurgy goes back more than two thousand years and physical material science—less than two ages. Against this background, the HEA conception with its child's 20-year age looks, at first glance, not entirely convincingly. Nevertheless, a publication boom on the HEA problem is observed in all economically developed countries.

The numerous overviews and monographs are published by leading scientists. It is shown that in the majority of cases the HEAs represent a single-phase solid solution with BCC-or FCC-lattice. The dominant factors in formation of structural phase states and properties were determined. The methods of manufacturing HEAs are being improved, in addition to the classic method of melt crystallisation the mechanical melting in ball mills in combination with electrospark and plasma smelting was developed. Thin films and coatings are produced by magnetron spraying. The melting under the action of low-energy electron beam or powerful ion beam may be considered to be a perspective method. Owing to the unique physico-mechanical properties (high wear resistance, increased strength, corrosion resistance, thermal stability, etc.) the fields of the HEAs application are extended.

In spite of a sufficient number of published papers, systematic studies on mechanical properties of alloys with a well-controlled and described microstructure are relatively little. In hope to find new properties of HEAs the authors sacrifice the necessity of thorough analysis of microstructural evolution in the process of loading.

It often leads to promising results which, unfortunately, may be considered only as preliminary ones. The situation is much better concerning the principle mechanisms of deformation and failure. These, in many respects, are similar to mechanisms of common alloys that allow one to look with optimism at perspectives for the development of new HEAs with improved properties. There are some complications due to multi-component nature of HEAs relating to atomic distribution. Is the distribution accidental or there is a short-range order? The connection between a short-range order and the mechanical properties of HEAs remains largely unclarified. In the future, the settlement of the problem of creating the HEAs with a high complex of properties will go in two directions. “From bottom to top”—by studies on little samples for determining perspective disadvantages with subsequent transfer to the use of massive products with control of micro- and macrostructure. The reverse approach “from top to bottom” is also justified—by employment of already effective HEAs and application of different mechanisms of strengthening to them for obtaining the properties that are considered superior to those of currently available alloys.

References

Egami, T., Guo, W., Rack, P.D., Nagase, T.: Metall. Mater. Trans. A **45**, 180 (2014)

Marik, S., Varghese, M., Sajilesh, K.P., Singh, D., Singh, R.P.: J. Alloys Compd. **769**, 1059 (2018)

Marik, S., Motla, K., Varghese, M., Sajilesh K.P., Singh, D., Breard, Y., Boullay, P., Singh, R.: **3**, 060602 (2019)

Meyers, M.A., Vöhringer, O., Lubarda, V.A.: Acta Mater. **49**, 4025 (2001)

Mishra, R.K., Shahi, R.: J. Alloys Compd. **821**, 153534 (2020)

Murtu, B.S., Yeh, J.W., Ranganathan, S.: High-Entropy Alloys, 1st edn. Elsevier, Butterworth-Heinemann (2014)

Pogrebnyak, A.D., Bagdasapyan, A.A., Yakushchenko, I.V., Beresnev, V.M.: Russ. Cheem. Rev. **83**, 1027 (2014)

Rogachev, A.S.: Phys. Met. Mater. Sci. **121**, 807 (2020)

Shaisultanov, D.G.: Structure and Mechanical Properties of CoCrFeNiX System (X = Mn, V, Mn and V, Al and Cu) High-Entropy Alloys (2015)

Shkodich, N.F., Spasova, M., Farle, M., Kovalev, D.Y., Nepapushev, A.A., Kuskov, K.V., Vergunova, Y.S., Scheck, Y.B., Rogachev, A.S.: J. Alloys Compd. **816**, 152611 (2020)

Sogabe, R., Goto, Y., Mizuguchi, Y.: **11**, 053102 (2018)

Vrtnik, S., Koželj, P., Meden, A., Maiti, S., Steurer, W., Feuerbacher, M., Dolinšek, J.: J. Alloys Compd. **695**, 3530 (2017)

Zhang, Y.: High-Entropy Materials. Springer Singapore, Singapore (2019)

Zuo, T., Gao, M.C., Ouyang, L., Yang, X., Cheng, Y., Feng, R., Chen, S., Liaw, P.K., Hawk, J.A., Zhang, Y.: Acta Mater. **130**, 10 (2017)

Chapter 5
Prediction of Phase Composition of Al-Co-Cr-Fe-Ni System High-Entropy Alloy

Abstract In this chapter a result solid solution formation of the high-entropy alloy was calculated according to Hume-Rothery rules. Graphical plots of entropy of mixing, enthalpy of mixing, differences in Pauling and Allen electronegativity as a function of the content of each element of a high-entropy alloy of Al-Co-Cr-Fe-Ni system have been built. The program HEApredict_v.1 for calculation of thermodynamic and phenomenological parameters as well as prediction of phase composition of quinary high-entropy alloys was described. The Research is supported by the Russian Science Foundation (project No. 20-19-00452).

Prediction of high-entropy alloys' (HEA) structure supposes a search for compositions forming disordered or ordered solid solutions that might be a base of alloy and then of alloying elements which would result in its strengthening and formation of strengthening phases in it. Unfortunately, a large number of HEAs possible compositions complicates the problem of search for the best, from the point of view of the required properties, composition of alloy. However, there are techniques making it possible to assess the formation of various structures and physical and mechanical properties corresponding to them using phenomenological criteria based on Hume-Rothery rules and/or thermodynamic parameters (Poletti and Battezzati 2014).

The main objective of analytical research on the calculation of atomic fractions of components forming the Al-Co-Cr-Fe-Ni system high-entropy alloy is a search for the optimal chemical composition of initial wires and their diameters that allow the production of alloy with prescribed microstructure and properties. The objective is attained by determination of effect of content of each of the elements of the Al-Co-Cr-Fe-Ni system on the formation of one or another phase composition in the range of change in atomic fraction of each component, x from 0 to 5. The selected range for making calculations and plotting graphs of dependences of different thermodynamic parameters on atomic fractions of components included both the areas having already been studied earlier and expanded the known data with a new prediction.

In order to carry out the analytical treatment, a large body of modern scientific papers of Russian and foreign authors consisting of 160 published works was analysed including information on effect of chemical elemental content of Al-Co-Cr-Fe-Ni-system on its microstructure, phase composition, mechanical and other

V. E. Gromov et al., *Structure and Properties of High-Entropy Alloys*, Advanced Structured Materials 107, https://doi.org/10.1007/978-3-030-78364-8_5

properties. It permitted a comparison of the obtained results with practical data of previous studies to be made and to demonstrate the reliability of calculation used for determination such phases as BCC, FCC, topologically close-packed, Laves phases and σ-phases.

For automatisation of calculations, the computer program written in "Visual Basic for Applications" programming language was developed. The program makes it possible to calculate thermodynamic and phenomenological parameters for quinary high-entropy alloys based on three databases with mechanical, physical and thermodynamic properties of more than 70 elements. The input data of the program are the names of chemical elements as well as their atomic fraction. Program interface, table of physical and electric properties of chemical elements as well as the table of enthalpy of mixing based on Miedema macroscopic model are located on separate sheets of paper in a book in Microsoft Excel medium. The output program data are values of chemical composition, entropy and enthalpy of mixing, the difference in radii of elements, melting temperature in Celsius and Kelvin degrees, thermodynamic parameter Ω permitting one to assess the formation of solid solution in high-entropy alloys, the concentration of valence electrons, Pauling and Allen electronegativity. The obtained values are used for the prediction of phase composition by their comparison with boundary conditions of formation of solid solution, Laves phases, BCC-, FCC- and topologically close-packed crystal lattice as well as σ-phase known from literature sources.

In order to choose the optimal chemical composition and diameter of initial wires based on the calculation of atomic fractions of components a formula whose detailed description is presented in Appendix was proposed. It was established that chemical composition of final material calculated theoretically corresponds to that determined practically.

In accordance with the purpose the following problems were formulated and solved:

1. Calculation of thermodynamic parameters and comparison of them with known criteria of phase formation.
2. Development of programme for calculation of thermodynamic and phenomenological parameters as well as prediction of phase composition of quinary high-entropy alloys.
3. Determination of the chemical composition of each strand of a stranded wire corresponding to the required chemical composition of final high-entropy alloy fabricated by means of wire-arc additive manufacturing.

5.1 Calculation of Thermodynamic Parameters and Comparison of Them with Known Criteria of Phase Formation

According to the Hume-Rothery rule, there are two factors that might affect the formation of solid solution in alloys. The first factor is the dimensional effects of component atoms. For alloys whose difference in atomic radii exceeds 15%, the formation of substitutional solid solution is the most improbable. The second factor is the chemical compatibility of components, that is difference in electronegativity or enthalpy of mixing. The larger is the difference in electronegativity (or the less is the enthalpy of mixing) the more probable that components of alloy make up the chemical compounds rather than solid solutions (Zhang et al. 2008).

In high-entropy alloys there are much more base components than in conventional alloys therefore constituent atoms have equal probability to occupy places in the crystal lattice with solid solution formation. In this connection, each atom of HEAs element may be considered as an atom of dissolved substance. In this case, solid solution structure has considerable distortions of crystal lattice caused by a substantial difference in atomic radii between so many components that differs HEA structure from structure of pure metals and conventional alloys. Therefore the Hume-Rothery modified rules are used to predict the formation of substitutional disordered solid solutions. It was demonstrated that solid solutions in HEAs are formed at $-20 \leq \Delta H_{mix} \leq 5$ kJ/mol, $1 < \delta_r < 6.6\%$ (Zhang et al. 2008) where ΔH_{mix} is enthalpy of mixing, and δ_r is difference in atomic radii of alloy components that are calculated from the following formulas:

$$\Delta H_{mix} = -R \sum 4\omega_{ij} c_i c_j$$

$$\omega_{ij} = \sum_{k=0}^{3} \Omega_k \left[(c_i - c_j)/(c_i + c_j) \right]^k$$

where ΔH_{mix} is enthalpy of mixing; ω_{ij} is parameter depending on concentration and characterising interaction between elements in solid solution.

$$\delta_r = 100\% \sqrt{\sum c_i (1 - r_i/\bar{r})^2}$$

where δ_r is difference in atomic radii of components; r_i is atomic radius of i-component; $\bar{r} = \sum_{i=1}^{n} c_i r_i$—is mean atomic radius.

In the research (Yang and Zhang 2012) the thermodynamic parameter $\Omega > 1.1$, which may also be used to predict the formation of solid solution for many high-entropy alloys, was suggested. The introduction of the parameter is caused by the contribution of entropy of mixing in the formation of solid solution:

$$\Omega = \frac{T_{melt} \Delta S_{mix}}{|\Delta H_{mix}|}$$

where Ω is thermodynamic parameter assessing a formation of solid solutions and intermetallic phases; ΔS_{mix} is entropy of mixing; T_{melt} is average melting temperature of alloy.

$$\Delta S_{mix} = -R \sum c_i ln c_i$$

where ΔS_{mix} is entropy of mixing, R is gas constant and c_i is atomic concentration of i—element.

$$T_{melt} = \sum_{i=1}^{n} c_i (T_{melt})_i$$

where $(T_{melt})_i$ is melting temperature of alloy of i-element.

Figures 5.1, 5.2 and 5.3 show the graphs displaying dependences of such parameters as Ω, ΔH_{mix} and δ_r on atomic fractions of high-entropy alloys components. The obtained results show that the crystal structure of solid solution will form in the entire range x = 0–5 in each Al-Co-Cr-Fe–Ni with the exception of $Al_xCoCrFeNi$ at x > 1.9.

Calculation of configurational entropy of mixing for high-entropy quinary alloy was done as a function of the content of atomic fraction of each component and is shown in Fig. 5.4.

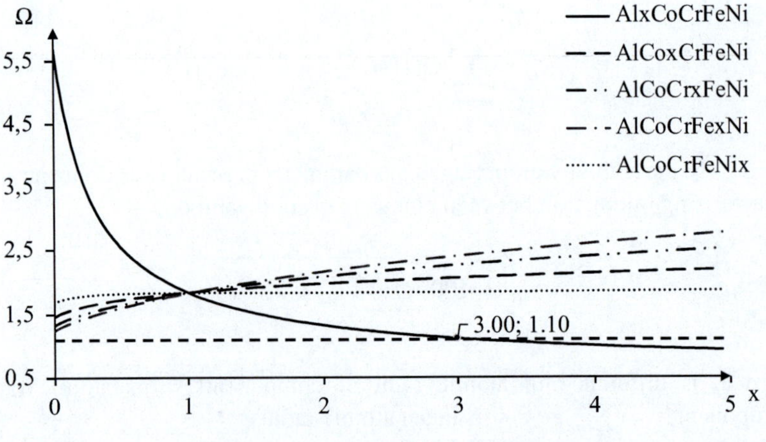

Fig. 5.1 A plot of the Ω parameter of Al-Co-Cr-Fe-Ni system HEA as a function of different content of each element. Boundary of solid solution formation is designated by the dotted line on the plot

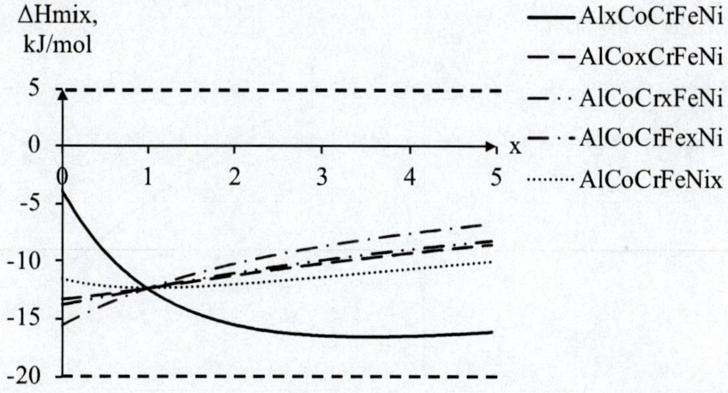

Fig. 5.2 A plot of the enthalpy of mixing ΔH_{mix} of Al-Co-Cr-Fe-Ni system HEA as a function of different content of each element. The upper and lower limits of solid solution formation are designated by dotted lines

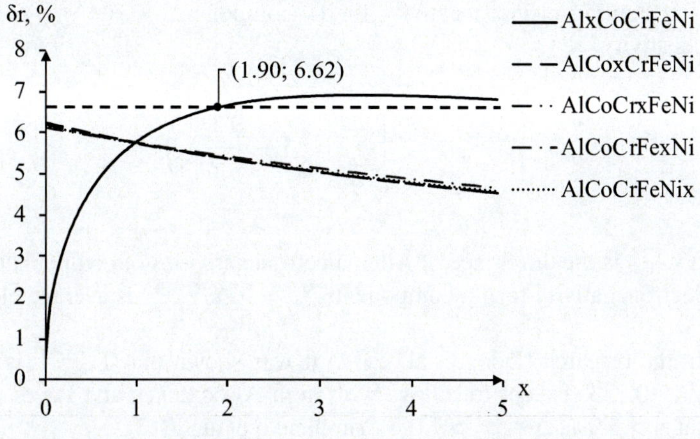

Fig. 5.3 A plot of the enthalpy of mixing δ_r of Al-Co-Cr-Fe-Ni system HEA as a function of different content of each element. The upper and lower limits of solid solution formation are designated by dotted lines

It is seen from Fig. 5.4 that the largest entropy of mixing is reached at equimolar relation of high-entropy alloy components.

Differences in electronegativity of $\Delta\chi_{\text{Pauling}}$ and $\Delta\chi_{\text{Allen}}$ allow one to assess the possibility of formation of the topologically close-packed crystal lattice (TCPCL) and Laves phases in high-entropy alloys:

$$\Delta\chi_{\text{Pauling}} = \sqrt{\sum_{i=1}^{n} c_i\left(\chi_i^{\text{Pauling}} - \overline{\chi}\right)^2}$$

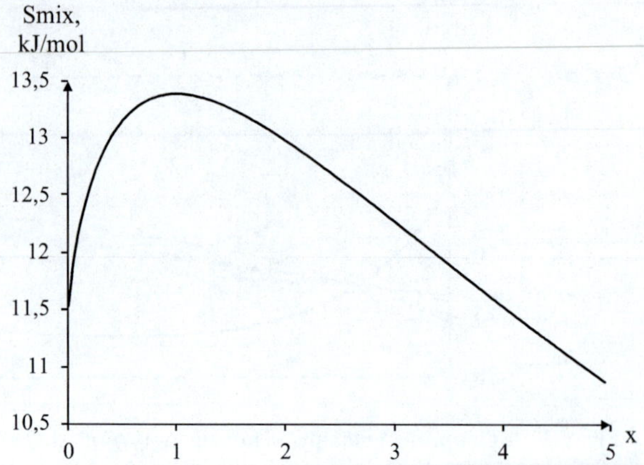

Fig. 5.4 A plot of the entropy of mixing ΔS_{mix} of Al-Co-Cr-Fe-Ni system HEA as a function of different content of each element

where $\Delta\chi_{Pauling}$ is the difference in electronegativities of components by Pauling; $\chi_i^{Pauling}$ is Pauling electronegativity for i—component, $\overline{\chi} = \sum c_i \chi_i$ is average electronegativity.

$$\Delta\chi_{\text{Allen}} = \sqrt{\sum_{i=1}^{n} c_i \left(1 - \frac{\chi_i^{\text{Allen}}}{\chi_a}\right)^2}$$

where $\Delta\chi_{\text{Allen}}$ is the difference of Allen electronegativities of components; χ_i^{Allen} is Allen electronegativity for I—component; $\chi_a = \sum c_i \chi_i^{\text{Allen}}$ is average electronegativity.

So, in the research (Dong et al. 2014) it was shown that TCPCL is formed at $\Delta\chi_{\text{Pauling}} > 0.133$ (except for alloys with high Al content), and Laves phases are formed at $\delta_r > 5.0\%$, $\Delta\chi_{\text{Allen}} > 7.0\%$ (Yurchenko et al. 2017).

The calculated parameter of electronegativity $\Delta\chi_{\text{Allen}}$ showed that Laves phases not observed in the given limits of X of high-entropy alloys of Al-Co-Cr-Fe–Ni system (Fig. 5.5).

The formation of topologically close-packed (TCP) crystal lattice, on the basis of calculations presented in Fig. 5.6, is possible in $AlCoCrFe_xNi$ ($x < 0.1$) alloy.

Determination of valence electron concentration (VEC) permits one to assess a possibility of BCC- and/or FCC-lattices formation in high-entropy alloys. In previous investigations it was demonstrated that BCC-lattice is formed at VEC < 6.87, FCC-lattice is formed at VEC ≥ 8 and BCC + FCC lattice is formed at 6.87 ≤ VEC < 8 (Guo et al. 2011). The σ-phase formation is also determined by VEC value provided that Cr and/or V are contained in the alloy: 6.88 ≤ VEC ≤ 7.84 (Tsai et al. 2013).

The VEC may be calculated by the following formula:

$$\text{VEC} = \sum c_i \text{VEC}_i$$

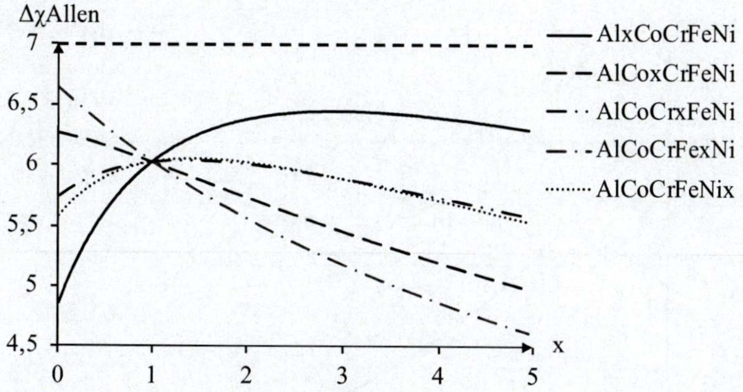

Fig. 5.5 A plot of Allen electronegativity as a function of the content of each element of Al-Co-Cr-Fe-Ni system HEA

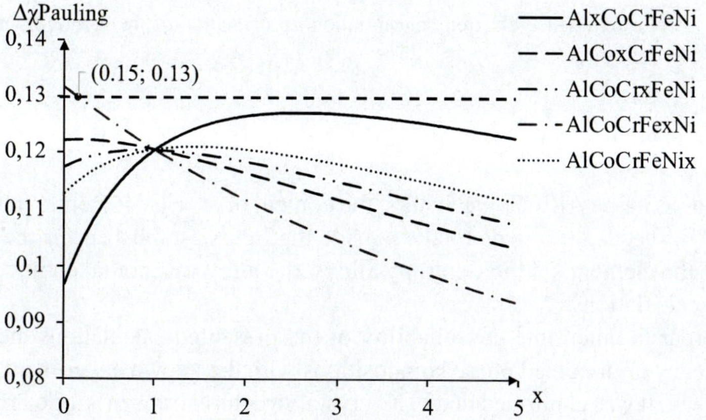

Fig. 5.6 A plot of Pauling electronegativity as a function of each element of Al-Co-Cr-Fe-Ni system HEA

where VEC is valence electron concentration; $(VEC)_i$ is valence electron concentration of i—element.

On the basis of the analysis of previous literature, it was established that alloys of Al-Co-Cr-Fe–Ni system having BCC lattice possess higher strength and hardness than those with FCC lattice. Therefore, the search was done by the criterion of BCC-lattice presence in the alloy being predicted.

Figure 5.7 presents the calculated curves of changes in valence electron concentration as a function of atomic fraction of HEA components. As is seen from the plots, the BCC-lattice is formed in $Al_xCoCrFeNi$ (x > 1.4), $AlCo_xCrFeNi$ (x = 0–0.2), $AlCoCr_xFeNi$ (x > = 2.9), $AlCoCrFeNi_x$ (x = 0–0.45) alloys, and σ-phase is formed in alloys $Al_xCoCrFeNi$ at x = 0.35–1.4, $AlCo_xCrFeNi$ at x = 0.25–3.75,

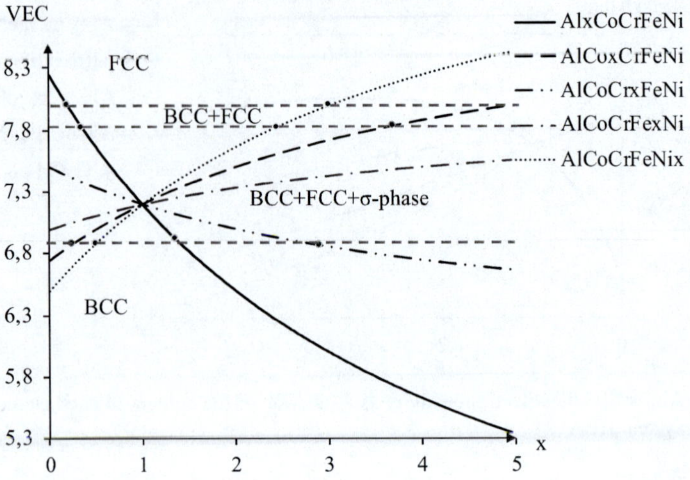

Fig. 5.7 A plot of valence electron concentration as a function of the atomic fraction of HEA components

$AlCoCr_xFeNi$ at x = 0–2.85, $AlCoCrFe_xNi$ x = 0–5, $AlCoCrFeNi_x$ at 0, 45 < x < 2.48.

As far as in the $AlCoCr_xFeNi$ alloy the content of at.% Cr at atomic fraction value x > 2.75 exceeds 40%, which falls outside the limits established for the content of each of the elements of high-entropy alloys, the alloy was not taken into account in further calculations.

In order to determine the reliability of the presented calculations the check for consistency of predicted phase compositions with the known experimental data was carried out. It was confirmed that FCC-crystal structure forms in $Al_xCoCrFeNi$ alloys at $0 \leq \times 0.4$ (Wang et al. 2012). The BCC + FCC—crystal structure was observed et al. $0.5 \leq x \leq 0.9$, which corresponds to predicted phase composition and boundaries of formation of single-phase BCC crystal lattice are extended as compared to the calculated ones and have values of x > 0.9.

According to experimental data (Kang et al. 2018) the BCC + B2 crystal structure forms at $0.25 \leq x \leq 1$ in $AlCo_xCrFeNi$ alloy. The formation of BCC-phase was determined in theoretical calculations.

In literary sources, it is shown that in $AlCoCrFe_xNi$ alloys at x = 0.2–0.6 the main phases are BCC + B2, and chemical compound Cr_3Ni_2 is also precipitated (Chen et al. 2015). At $0.8 \leq x \leq 2.0$ Cr_3Ni_2 dissolves completely and the BCC + B2 crystal lattice is stabilized. When comparing the experimental data with those obtained theoretically it may be concluded that the formation of BCC-phase and σ-phase Cr_3Ni_2 was predicted successfully.

Practical results of study of $AlCoCrFeNi_x$ alloy phase composition showed that at x = 1.5 the BCC + FCC + B2 crystal structure is observed, and at $1.8 \leq x \leq 3.0$ the FCC + BCC (Cao et al. 2019) crystal structure. The data are consistent with those calculated theoretically.

Thus, the comparison of the results obtained theoretically with experimental data demonstrated the reliability of calculations used to determine such phases as BCC, FCC, TCP, Laves phases and σ-phases.

5.2 Program for Calculation of Thermodynamic Parameters and Prediction of Phase Composition of Quinary High-Entropy Alloys

The developed program for the calculation of thermodynamic parameters permitting the prediction of phase composition of quinary high-entropy alloys has a simple interface and is presented in Fig. 5.8.

Input data of the program are names of chemical elements as well as atomic fractions. Output data of the program are values of chemical composition, entropy and enthalpy of mixing, difference in atomic radii of elements, melting temperature in degrees Celsius and Kelvin, thermodynamic parameter Ω permitting one to assess a formation of solid solution in high-entropy alloys, valence electron concentration, Pauling and Allen electronegativity. The obtained values are used for predicting

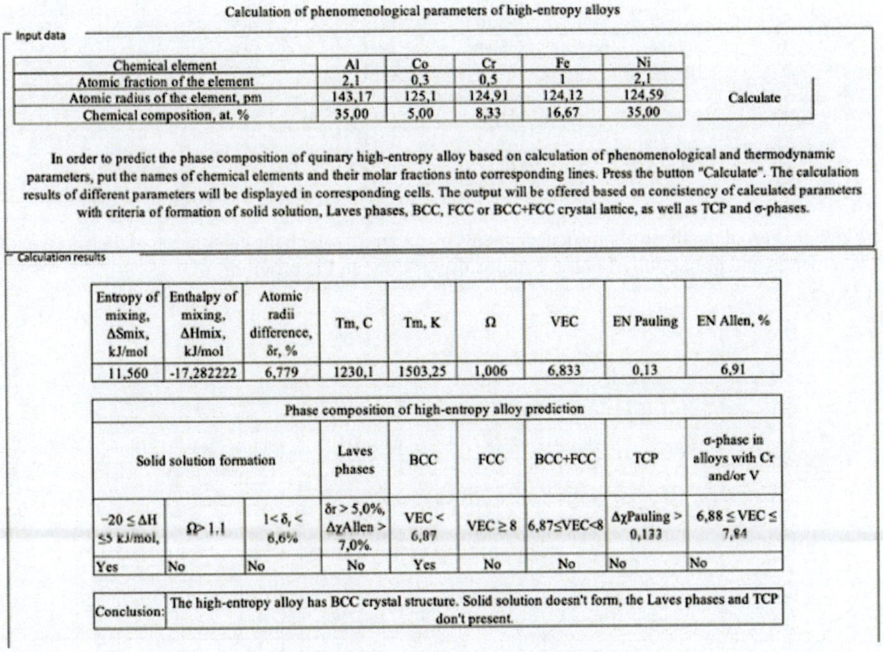

Calculation of phenomenological parameters of high-entropy alloys

Input data

Chemical element	Al	Co	Cr	Fe	Ni
Atomic fraction of the element	2,1	0,3	0,5	1	2,1
Atomic radius of the element, pm	143,17	125,1	124,91	124,12	124,59
Chemical composition, at. %	35,00	5,00	8,33	16,67	35,00

Calculate

In order to predict the phase composition of quinary high-entropy alloy based on calculation of phenomenological and thermodynamic parameters, put the names of chemical elements and their molar fractions into corresponding lines. Press the button "Calculate". The calculation results of different parameters will be displayed in corresponding cells. The output will be offered based on concistency of calculated parameters with criteria of formation of solid solution, Laves phases, BCC, FCC or BCC+FCC crystal lattice, as well as TCP and σ-phases.

Calculation results

Entropy of mixing, ΔSmix, kJ/mol	Enthalpy of mixing, ΔHmix, kJ/mol	Atomic radii difference, δr, %	Tm, C	Tm, K	Ω	VEC	EN Pauling	EN Allen, %
11,560	-17,282222	6,779	1230,1	1503,25	1,006	6,833	0,13	6,91

Phase composition of high-entropy alloy prediction								
Solid solution formation			Laves phases	BCC	FCC	BCC+FCC	TCP	σ-phase in alloys with Cr and/or V
−20 ≤ ΔH ≤5 kJ/mol,	Ω> 1,1	1< δ, < 6,6%	δr > 5,0%, ΔχAllen > 7,0%.	VEC < 6,87	VEC ≥ 8	6,87≤VEC<8	ΔχPauling > 0,133	6,88 ≤ VEC ≤ 7,84
Yes	No	No	No	Yes	No	No	No	No

Conclusion:	The high-entropy alloy has BCC crystal structure. Solid solution doesn't form, the Laves phases and TCP don't present.

Fig. 5.8 Interface of program for calculation of thermodynamic parameters and prediction of phase composition of quinary high-entropy alloys

the phase composition by their comparison with boundary conditions of solid solution formation, Laves phases, BCC-, FCC- and/or topologically close-packed crystal lattices as well as σ-phase known from literary sources.

Table of physical and electrical properties of chemical elements (Fig. 5.9) as well as Table of values of enthalpies of mixing based on Miedema macroscopic model (Takeuchi and Inoue 2016) are placed on separate worksheets in the book in Microsoft Excel (Fig. 5.10).

	A	B	C	D	E	F	G
1	Element	Atomic number	Atom radius (pm)	Tm (K)	Pauling EN	VEC	Allen EN
2	H	1	25,00				2,3
3	Li	3	151,94	454	0,98	1	0,91
4	Be	4	112,80	1560	1,57	2	1,58
5	B	5	82,00	2348	2,04	3	2,05
6	C	6	77,30	4742	2,55	4	2,54
7	N	7	75,00	63	3,04	5	3,07
8	O	8	73,00	54	136,26	137	3,61
9	Na	11	185,70	371	0,93	1	0,87
10	Mg	12	160,13	923	1,31	2	1,29
11	Al	13	143,17	933	1,61	3	1,61
12	Si	14	115,30	1687	1,90	4	1,92
13	P	15	106,00	317	2,19	5	2,25
14	S	16	102,00	388	2,58	6	2,59
15	K	19	231,00	337	0,82	1	0,73
16	Ca	20	197,60	1115	1,00	2	1,03
17	Sc	21	164,10	1814	1,36	3	1,19
18	Ti	22	146,15	1941	1,54	4	1,38
19	V	23	131,60	2183	1,63	5	1,53
20	Cr	24	124,91	2180	1,66	6	1,65
21	Mn	25	135,00	1519	1,55	7	1,75
22	Fe	26	124,12	1811	1,83	8	1,80
23	Co	27	125,10	1768	1,88	9	1,84
24	Ni	28	124,59	1728	1,91	10	1,88

Fig. 5.9 Part of database of physical elements by means of which the calculation of thermodynamic parameters and prediction of phase composition is done in the program

	B	C	D	E	F	G	H	I	J	K	L	M	N	O	P	Q	R	S	T	U	V	W
1		1	3	4	5	6	7	11	12	13	14	15	19	20	21	22	23	24	25	26	27	28
2		H	Li	Be	B	C	N	Na	Mg	Al	Si	P	K	Ca	Sc	Ti	V	Cr	Mn	Fe	Co	Ni
3	H		-25	2	5	-3	-18	-16	-19	-8	-19	1	-16	-46	-60	-54	-39	-28	-34	-23	-23	-23
4	Li	-25		-5	-6	-61	-145	4	0	-4	-30	-46,	11	-1	12	34	37	35	19	26	8	1
5	Be	2	-5		0	-15	-39	18	-3	0	-15	-4	27	-14	-36	-30	-16	-7	-10	-4	-4	-4
6	B	5	-6	0		-10	-28	18	-4	0	-14	5	27	-22	-55	-58	-42	-31	-32	-26	-24	-24
7	C	-3	-61	-15	-10		-2	-45	-55	-36	-39	-5,	-43	-89	-118	-109	-82	-61	-66	-50	-42	-39
8	N	-18	-145	-39	-28	-2		-141	-134	-92	-81	-25,	-152	-201	-224	-190	-143	-107	-119	-87	-75	-69
9	Na	-16	4	18	18	-45	-141		10	13	-11	-27,	1	1	34	68	73	71	49	62	41	32
10	Mg	-19	0	-3	-4	-55	-134	10		-2	-26	-40,	20	-6	-3	16	23	24	10	18	3	-4
11	Al	-8	-4	0	0	-36	-92	13	-2		-19	-21,	23	-20	-38	-30	-16	-10	-19	-11	-19	-22
12	Si	-19	-30	-15	-14	-39	-81	-11	-26	-19		-26,	-4	-51	-74	-66	-48	-37	-45	-35	-38	-40
13	P	1	-46,	-4	5	-5,	-25,	-27,	-40,	-21,	-26,		-25,	-82	-112,5	-100,5	-71,	-50	-58	-40	-36	-35
14	K	-16	11	27	27	-43	-152	1	20	23	-4	-25,		12	58	94	96	91	66	81	55	45
15	Ca	-46	-1	-14	-22	-89	-201	1	-6	-20	-51	-82	12		17	43	44	38	19	25	2	-7
16	Sc	-60	12	-36	-55	-118	-224	34	-3	-38	-74	-113	58	17		8	7	1	-8	-11	-30	-39
17	Ti	-54	34	-30	-58	-109	-190	68	16	-30	-66	-101	94	43	8		-2	-7	-8	-17	-28	-35
18	V	-39	37	-16	-42	-82	-143	73	23	-16	-48	-71,	96	44	7	-2		-2	-1	-7	-14	-18
19	Cr	-28	35	-7	-31	-61	-107	71	24	-10	-37	-50	91	38	1	-7	-2		2	-1	-4	-7
20	Mn	-34	19	-10	-32	-66	-119	49	10	-19	-45	-58	66	19	-8	-8	-1	2		0	-5	-8
21	Fe	-23	26	-4	-26	-50	-87	62	18	-11	-35	-40	81	25	-11	-17	-7	-1	0		-1	-2
22	Co	-23	8	-4	-24	-42	-75	41	3	-19	-38	-36	55	2	-30	-28	-14	-4	-5	-1		0
23	Ni	-23	1	-4	-24	-39	-69	32	-4	-22	-40	-35	45	-7	-39	-35	-18	-7	-8	-2	0	

Fig. 5.10 Part of database of values of enthalpies of mixing based on Miedema macroscopic model. The values used in the research are highlighted with colour

The program developed for the computer is registered in Federal Institute of Industrial Property (FIIP): 2,020,666,726 "Program HEApredict_v.1 for calculation of thermodynamic criteria and prediction of phase composition of quinary high-entropy alloys." The authors are Osintsev K.A., Konovalov S.V., Panchenko I.A., Gromov V.E. Copyright owner is Federal State Budjet Educational Institution of Higher Education "Siberian State Industrial University" (RU). Date of publication and bulletin number: 15.12.2020 Bul. No. 12.

5.3 Determination of the Chemical Composition of a Stranded Wire Corresponding to the Required Chemical Composition of Final High-Entropy Alloy Fabricated by Wire-Arc Additive Manufacturing

In order to choose the optimal chemical composition and diameter of initial wires, it was necessary to have a possibility to predict a final chemical composition as a function of these parameters. To allow the assessment of the chemical composition of sample deposited in layer-by-layer fashion it was suggested to consider n-component wire as a single-strand wire. For it, the total chemical composition of its cross-section was calculated.

Consider, as a model, a cross cut from wire consisting of three strands (1, 2, and 3), with diameter d_1, d_2 and d_3 and 1 atom in thickness. Assume that each strand has a continuous surface (without defects) and atoms are at infinitely small distances from each other, thus occupying the whole area of cross-section. Then the area of surface occupied by atoms will be proportional to their chemical composition in initial strands. If we know the atomic relations of elements of each of initial strands, then, in order to find total chemical composition, it is necessary to divide number of atoms of each of the elements into total number of atoms. Thus, calculation is reduced to determination of total amount of atoms of each of elements that depend on chemical composition of initial wires and their diameters.

Consider calculation of atomic per cent of "a" element in strand 1. If "a" element is present only in one of strands, then number of atoms is determined as

$$N_{\mathrm{a}} = n_{\mathrm{a}} \cdot N_A = \frac{m_a}{M_a} \cdot N_A = \frac{\rho_a \cdot V_1}{M_a} \cdot N_A$$

where N_a is amount of atoms of "a" element, N_A is Avogadro number, n_a is amount of moles of "a" element, m_a is atomic mass of "a" element, M_a is molar mass of "a" element, ρ_a is density of "a" element, V_1 is volume occupied by "a" element atoms in a strand.

As the thickness of strand cut being considered is one atom, take a fraction of strand cross-sectional area for V_1 being occupied by atoms. The fraction is equal to the product of total area of S_a strand by chemical composition of initial wire expressed in fractions, x_a then:

$$N_{\mathrm{a}} = \frac{\rho_a \cdot S_a}{M_a} \cdot N_A = \frac{\rho_a \cdot pi \cdot D_1^2 \cdot x_{1a}}{M_a \cdot 4} \cdot N_A$$

where S_a is cross-section area of strand 1, D_1 is diameter of strand 1, x_{1a} is atomic fraction of "a" element in strand 1.

If "a" element is present in several strands, then the total amount of atoms we find by summation of amount of atoms of "a" element in each of strands. For example, if atoms of "a" element are present in two strands, then its total number of atoms equals to:

$$N_{\mathrm{a}} = \frac{\rho_a \cdot pi \cdot D_1^2 \cdot x_{1a}}{M_a \cdot 4} \cdot N_A + \frac{\rho_a \cdot pi \cdot D_2^2 \cdot x_{2a}}{M_a \cdot 4} \cdot N_A$$

where D_2 is diameter of strand 2, and x_{2a} is atomic fraction of "a" element in strand 2.

Thus, having calculated the amount of atoms of each of elements, being a part of the composition of strands of stranded wire, the atomic per cent of each of elements can be found by the following formula:

$$\mathrm{A} = \frac{N_{\mathrm{a}}}{N_{\mathrm{total}}} \cdot 100\% = \frac{N_{\mathrm{a}}}{\mathrm{N}_a + \mathrm{N}_b + \mathrm{N}_c + \mathrm{N}_d + \mathrm{N}_e} \cdot 100\%,$$

where $N_{total} = N_a + N_b + N_c + N_d + N_e$ is total amount of atoms in entire cross-cut of stranded wire.

Using the formula the optimum chemical compositions of strands for multi-stranded wire were calculated and diameters corresponding to them were found: aluminium wire A995 (Al ≈ 99.95%, diameter 0.5 mm), chromium-nickel wire $Cr_{20}Ni_{80}$ (Cr ≈ 20%, Ni ≈ 80%, diameter 0.4 mm) and wire made of precision alloy 29NK (Co ≈ 17%, Fe ≈ 54%, Ni ≈ 29%, diameter 0.4 mm). The calculated elemental composition expressed in atomic per cent, when the given wires are used as initial material, is represented by: Al—34.92 at. %, Co—6.69 at. %, Cr—5.56 at. %, Fe—17.17 at. %, Ni—35.65 at. %. The values of thermodynamic parameters of obtained alloy are shown in Tables 5.1, 5.2 and 5.3. As can be seen from the results of

Table 5.1 Input data for calculation of thermodynamic and phenomenological parameters and prediction of phase composition

Chemical element	Al	Co	Cr	Fe	Ni
Atomic fraction of the element	2.1	0.3	0.5	1	2.1
Atomic radius of the element, pm	143.17	125.1	124.91	124.12	124.59
Chemical composition, at. %	35.00	5.00	8.33	16.67	35.00

Table 5.2 Calculated thermodynamic and phenomenological parameters

Entropy of mixing	Enthalpy of mixing, ΔH, kJ/mol	Difference in atomic radii, δr, %	Tm, C	Tm, K	Ω	VEC	EN Pauling	EN Allen, %
11.560	−17.282	6.779	1230.1	1503.25	1.006	6.833	0.13	6.91

prediction of phase composition presented in Table 5.3 the given chemical composition forms no solid solution as well as assumes no formation of Laves phases, TCP phases and σ-phases but it has BCC—crystal lattice. Since the phase composition meets the criteria of search, then the wires for manufacturing the high-entropy alloy of Al-Co-Cr-Fe–Ni system were chosen.

As a result of preforming the problem to determine chemical composition of each of strands of stranded wire corresponding to the required chemical composition of final high-entropy alloy fabricated by wire-arc additive manufacturing the strands of the following elemental composition were selected: (Al ≈ 99.95%, diameter 0.5 mm), chromium-nickel strands Cr20Ni80 (Cr ≈ 20%, Ni ≈ 80%, diameter 0.4 mm) and strands made of precision alloy 29NK (Co ≈ 17%, Fe ≈ 54%, Ni ≈ 29%, diameter 0.4 mm). The predicted chemical composition of high-entropy alloy produced by wire-arc additive manufacturing using a multi-stranded wire of selected strands is the following: Al—34.92 at. %, Co—6.69 at. %, Cr—5.56 at. %, Fe—17.17 at. %, Ni—35.65 at. %.

5.4 Conclusion

As a result of the research the boundary conditions of solid solution formation of high-entropy alloy according to Hume-Rothery rules were calculated, graphical plots of entropy of mixing, enthalpy of mixing, differences in Pauling and Allen electronegativity as a function of the content of each element of high-entropy alloy of Al-Co-Cr-Fe–Ni system have been built. The program HEApredict_v.1 for calculation of thermodynamic and phenomenological parameters as well as prediction of phase composition of quinary high-entropy alloys has been developed. On the basis of calculation of thermodynamic parameters and comparison of them with the known criteria of phase formation following results have been obtained:

1. The BCC-phase forms in $Al_xCoCrFeNi$ at x > 1.45, $AlCo_xCrFeNi$ at x = 0–0.2, $AlCoCr_xFeNi$ at x > = 2.9, $AlCoCrFeNi_x$ at x = 0–0.45.
2. The FCC-phase forms in $Al_xCoCrFeNi$ at x = 0–0.15, $AlCoCrFeNi_x$ at x > 3.
3. The mixed crystal structure consisting of BCC and FCC crystal lattices forms in $Al_xCoCrFeNi$ at x = 0.2–1.4, $AlCo_xCrFeNi$ at x > 0.25, $AlCoCr_xFeNi$ at x = 0–2.85, $AlCoCrFe_xNi$ x = 0–5, $AlCoCrFeNi_x$ < 3.

Table 5.3 Results of prediction of phase composition based on calculated thermodynamic and phenomenological parameters

Prediction of phase composition of high-entropy alloy								
Formation of solid solution			Laves phases	BCC	FCC	BCC + FCC	TCP	σ-phase in alloys with Cr and/or V
$-20 \le \Delta H \le 5$ kJ/mol,	$\Omega > 1.1$	$1 < \delta r < 6{,}6\%$	$\delta r > 5.0\%$, $\Delta\chi Allen > 7.0\%$	$VEC < 6.87$	$VEC \ge 8$	$6{,}87 \le VEC < 8$	$\Delta\chi Pauling > 0.133$	$6.88 \le VEC \le 7.84$
Yes	No	No	No	Yes	No	No	No	No
Conclusion	Solid solution is not formed, Laves phases, σ-phases and TCP—phases are absent, there is BCC crystal lattice							

4. The formation of solid solution of crystal structure is observed in entire studied range of atomic fraction of components x = 0–5 of Al-Co-Cr-Fe–Ni system, except for $Al_xCoCrFeNi$ at $x > 1.9$.
5. The Laves phase is not observed in Al-Co-Cr-Fe–Ni systems with different atomic relation of elements.
6. The TCP-phase may be stable only at a small amount of Fe in the $AlCoCrFe_xNi$ ($x < 0.1$) alloy.
7. The σ-phase forms in the alloys $Al_xCoCrFeNi$ at x = 0.35–1.4, $AlCo_xCrFeNi$ at x = 0.25–3.75, $AlCoCr_xFeNi$ at x = 0–2.85, $AlCoCrFe_xNi$ x = 0–5, $AlCoCrFeNi_x$ at $0.45 < x < 2.48$.

References

Guo, S., Ng, C., Lu, J., Liu, C.T.: J. Appl. Phys. **109**, 103505 (2011)
Chen, Q., Zhou, K., Jiang, L., Lu, Y., Li, T.: Arab. J. Sci. Eng. **40**, 3657 (2015)
Cao, L., Wang, X., Wang, Y., Zhang, L., Yang, Y., Liu, F., Cui, Y.: Appl. Phys. A Mater. Sci. Process. **125**, 1 (2019)
Dong, Y., Lu, Y., Jiang, L., Wang, T., Li, T.: Intermetallics **52**, 105 (2014)
Kang, M., Lim, K.R., Won, J.W., Na, Y.S.: J. Alloys Compd. **769**, 808 (2018)
Poletti, M.G., Battezzati, L.: Acta Mater. **75**, 297 (2014)
Takeuchi, A., Inoue, A.: J. Postgrad. Med. Inst. **30**, 80 (2016)
Tsai, M.H., Tsai, K.Y., Tsai, C.W., Lee, C., Juan, C.C., Yeh, J.W.: Mater. Res. Lett. **1**, 207 (2013)
Wang, W.-R., Wang, W.-L., Wang, S.-C., Tsai, Y.-C., Lai, C.-H., Yeh, J.-W.: Intermetallics **26**, 44 (2012)
Yang, X., Zhang, Y.: Mater. Chem. Phys. **132**, 233 (2012)
Yurchenko, N., Stepanov, N., Salishchev, G.: Mater. Sci. Technol. (united Kingdom) **33**, 17 (2017)
Zhang, Y., Zhou, Y.J.Y.J., Lin, J.P.J.P., Chen, G.L.G.L., Liaw, P.K.P.K.: Adv. Eng. Mater. **10**, 534 (2008)

Chapter 6
High-Entropy Alloys of AlCoCrFeNi-System

Abstract In this chapter, the high-entropy alloy of Al-Co-Cr-Fe-Ni system was manufactured by the wire-arc additive manufacturing technology and irradiated with high-intensity electron beams. The as-manufactured alloy has high microhardness but a brittle structure. The compression tests have shown that the ultimate strength of the material manufactured by different modes can vary from 550 to 1899 MPa, which is correlated with the data obtained from the previous articles on the high-entropy alloy Al-Co-Cr-Fe-Ni system. It has been established that HEA irradiation by the pulsed electron beam is accompanied by the release of grain boundaries from the second phase precipitations that is indicative of material homogenisation. It has been revealed that irradiation of high-entropy alloy by intense pulsed electron beam results in an increase in strength and plasticity of the material. The ultimate compression strength increases 1.1–1.6 times. The largest value of ultimate compression strength of 2179 MPa were obtained in the alloy processed by an electron beam with an energy density of 30 J/cm^2. The Research is supported by the Russian Science Foundation (project No. 20-19-00452).

6.1 Technique of High-Entropy Alloy Formation

To form the bulk samples (Fig. 6.1a) of high-entropy alloy (HEA) of AlCoCrFeNi system a multi-component (stranded) wire was used as the initial material. It consisted of three strands of different elemental composition: aluminium wire (Al $\approx$ 99.95%, 0.5 mm in diameter), chromium-nickel wire $Cr_{20}Ni_{80}$ (Cr $\approx$ 20%, Ni $\approx$ 80%, 0.4 mm in diameter), as well as wire of precision alloy 29NK (Co $\approx$ 17%, Fe $\approx$ 54%, Ni $\approx$ 29%, 0.4 mm in diameter). Multi-component wire was obtained by automated wire stranding of the given three wires (Fig. 6.1b). The choice of wires of the grades and their diameter was dictated by the fact that they ensured the production of high-entropy alloy of a preliminary calculated composition.

Multi-component wire was manufactured by automated stranding of the given three wires on original setup. To choose the optimal mode of stranding providing a successful (without jamming and ruptures) passage of wire in directing channel and

V. E. Gromov et al., *Structure and Properties of High-Entropy Alloys*, Advanced Structured Materials 107, https://doi.org/10.1007/978-3-030-78364-8_6

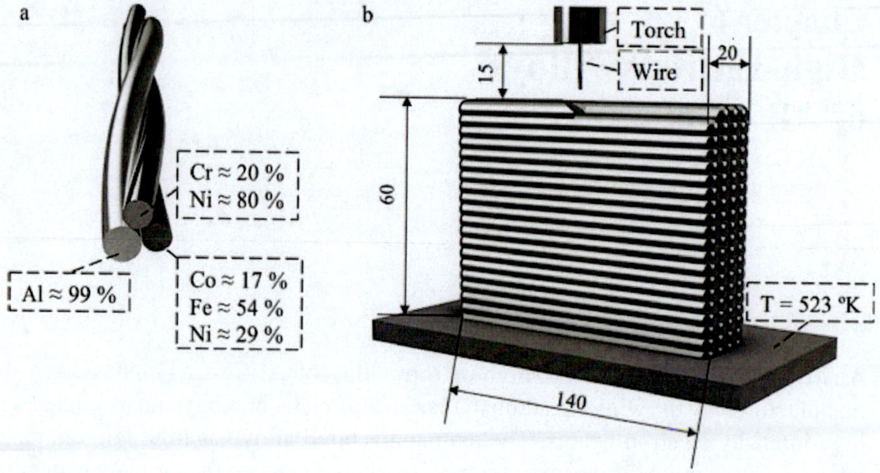

Fig. 6.1 **a** Diagram of wire-arc additive manufacturing using a cable consisting of 3 strands; **b** 3D-model of the wire used

torch, the rotation frequency of receiving coil and that of feeding coils vary. Diameter of combined wire manufactured by the method amounted to ≈ 1 mm.

Fabrication of HEA samples was done by layer-by-layer deposition on steel substrate by means of wire-arc additive manufacturing (WAAM) (cold metal transfer (CMT)) in atmosphere of inert gas (Ar ≈99.99%) (Fig. 6.1a). The following working parameters of deposition complex were constant: rate of wire feeding—8 m/min, deposition voltage—17 V, travel speed—0.3 m/min, gas supply speed (Ar)—14 l/min.

Three approaches were used in the formation of HEA samples: (1) deposition of all metal layers "from left to right," at substrate temperature of 25 °C; (2) deposition of all metal layers "from left to right" at substrate temperature of 250 °C; (3) deposition of metal from left to right then from right to left" and vice versa, substrate temperature of 250 °C (Fig. 6.2).

The resultant samples of HEA had sizes of ≈ 60 × 140 × 20 mm and were parallelepipeds consisting of 20 deposited layers in height and 4 layers in thickness. The X-ray fluorescence analysis showed that the manufactured alloy had the following

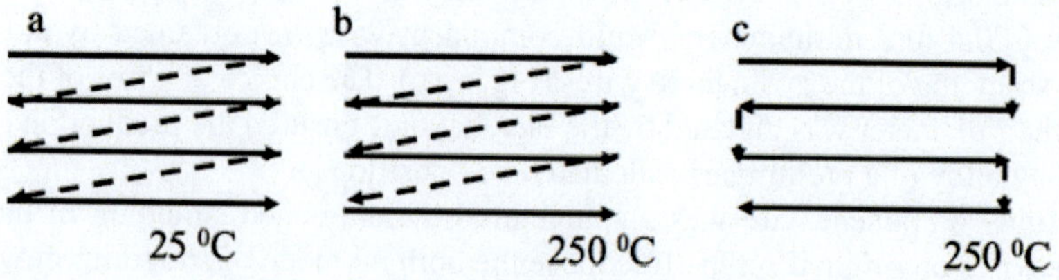

Fig. 6.2 **a** Deposition path of sample layers according to mode 1 **b** Deposition path of sample layers according to mode 2, **c** Deposition path of sample layers according to mode 3. Solid arrows indicate the direction of torch motion with arc switched on, dotted arrows—with arc switched off

elemental composition (wt. %): 15.64 Al, 7.78 Co, 8.87 Cr, 22.31 Fe, 44.57 Ni. The atoms of impurity elements (wt. %) 0.53 Si, 0.18 Cu, 0.098 Ti, being a part of initial wires were present in alloy composition as well. The resultant alloy corresponds to the non-equimolar high-entropy alloy of Al-Co-Cr-Fe-Ni system in its atomic composition.

6.2 Structure and Phase Composition of AlCoCrFeNi HEA Produced by Technology of Wire-Arc Additive Manufacturing

Elemental and phase composition of the alloy, state of the defective substructure were studied by the methods of scanning electron microscopy (microscope "LEO EVO 50", Carl Zeiss) with energy dispersion analyser an INCA-energy and TESCAN VEGA with energy dispersion analyser an INCAx-act) and transmission electron diffraction microscopy (instrument JEM 2100, JEOL). In order to reveal the alloy microstructure, the etching of samples preliminary polished on abrasive paper of different grain sizes and abrasive paste by HNO_3 and HCl solution in the ratio of 1:3 was performed. Phase composition and crystal lattice state of main phases of HEA samples were studied by the methods of X-ray phase and X-ray structural analysis (X-ray diffractometer Shimadzu XRD 6000 and DRON-7); the research was done in radiation Cu-Kα1. Phase composition analysis was performed using database PDF 4+ and programme of full-profile analysis POWDER CELL 2.4.

The images of microstructure of HEA sample cross-section allowed us to determine that deposited layers had a dendritic structure (Fig. 6.3).

Dendrites are oriented along the heat removal direction. Grain sizes vary in the limits from 4 μm to 15 μm and increase as the interface of deposited layers is approached.

A characteristic image of HEA samples' structure obtained by the methods of scanning electron microscopy is shown in Fig. 6.4. Surface etching of HEA samples results in revealing the grain structure (Fig. 6.4a). Grain sizes vary in the limits from 4 to 15 μm that is in agreement with the results obtained by the method of optical microscopy (Fig. 6.3). Along boundaries and in grain volume the second phase inclusions are revealed (Fig. 6.4b, inclusions are indicated by arrows).

Elemental composition of the alloy was studied by method of micro-X-ray spectral analysis. Figure 6.5a shows the energy spectra of the alloy section, the electron microscopic image of which is shown in Fig. 6.5b. It is clearly seen that the material under study contains aluminium, iron, nickel, chromium, and cobalt atoms. The results of the quantitative analysis of the elemental composition of the alloy under study (averaging was carried out over five randomly selected sections) showed that the main elements of the alloy section under study were aluminium (36.5 at. %), nickel (33.7 at. %), iron (16.4 at. %), chrome (8, 6 at. %) and cobalt (4.9 at.%).

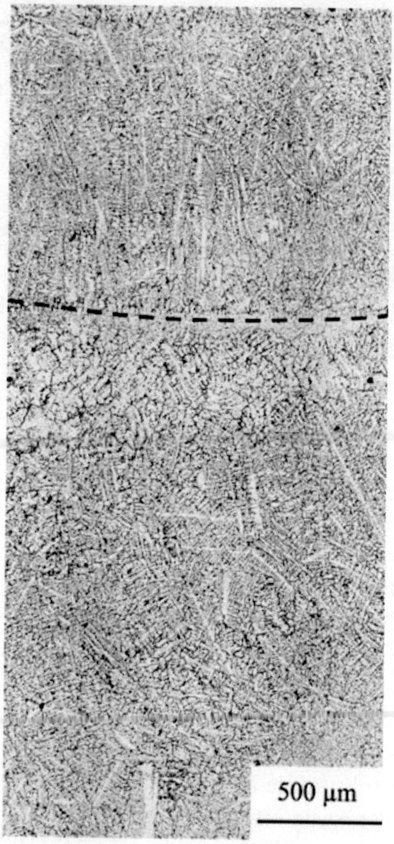

Fig. 6.3 Image of cross-section of HEA alloy sample. Dotted line indicates a boundary between deposited layers. Optical microscopy of etched metallographic section

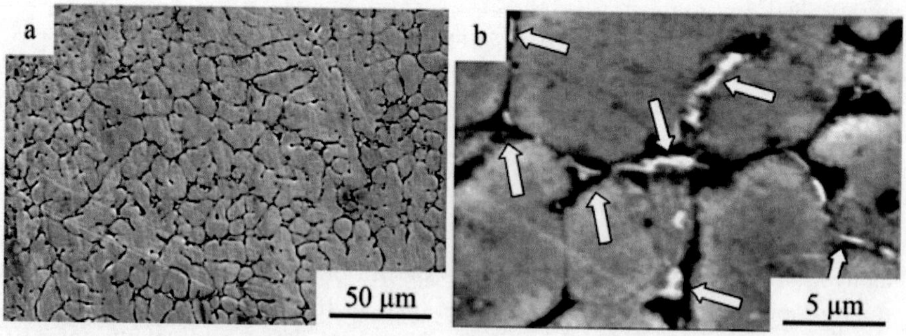

Fig. 6.4 Grain structure of HEA surface. The second-phase inclusions are indicated by arrows in **b**. Scanning electron microscopy

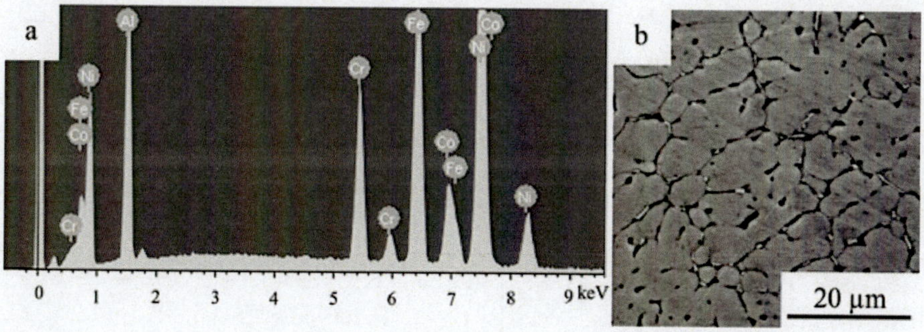

Fig. 6.5 Energy spectra (**a**) obtained from HEA were whose electron microscopic image is presented in (**b**)

By mapping methods, it is established that near-boundary volumes of the alloy (volumes located along grain boundaries) are enriched in chromium and iron atoms, the grain volume is enriched in nickel and aluminium atoms and cobalt is quasi-uniformly distributed in the alloy (Fig. 6.6).

Energy-dispersive X-ray spectroscopy of samples cross-section performed in every 5 mm allowed us to detect that the mentioned elements are distributed homogeneously along the entire material volume in the following composition: 35.67 ± 1.34 at. % Al, 33.79 ± 0.46 at. % Ni, 17.28 ± 1.83 at. % Fe, 8.28 ± 0.15 at. % Cr and

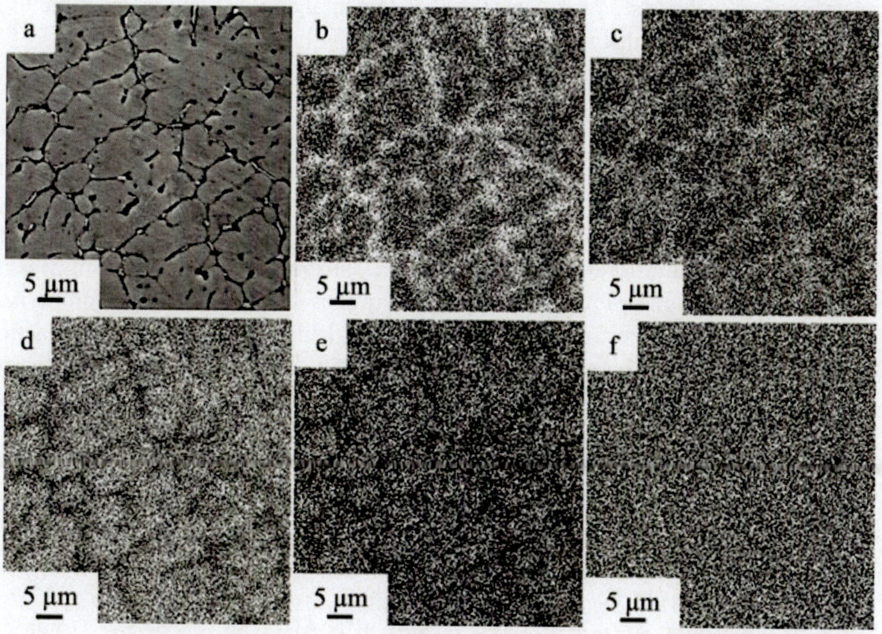

Fig. 6.6 SEM (**a**) of HEA surface and images of the given surface portion obtained in characteristic X-ray radiation of chromium (**b**), iron (**c**), nickel (**d**), aluminium (**e**) and cobalt (**f**) atoms

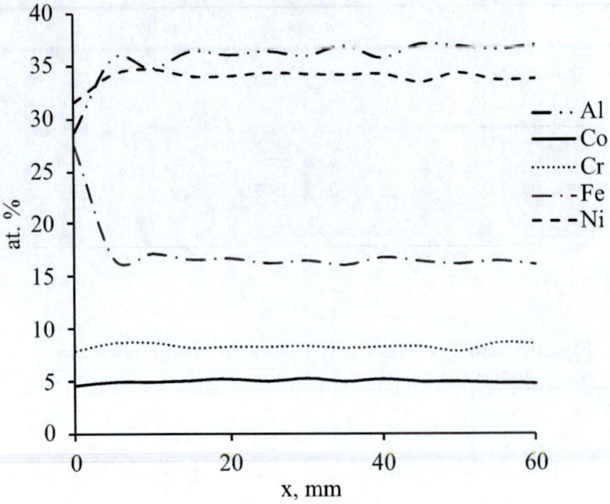

Fig. 6.7 Change in the content of chemical elements depending on the distance from the substrate

4.99 ± 0.09 at. % Co. However, the increase in iron content relative to its content in material volume was revealed at the boundary with the substrate. It may be caused by the fact that in the process of wire-arc additive manufacturing a mixing of deposited metal and base material takes place. As iron content is not detected at a distance of 5 mm from the conventional boundary with substrate and value of deposited layer reaches 3 mm the conclusion may be made that substrate affects only the chemical composition of the first layer. A plot of the elemental composition of manufactured alloy as a function of distance from the substrate is shown in Fig. 6.7.

Thermogravimetric analysis allowed us to determine the melting temperature of the produced high-entropy alloy (Fig. 6.8). Differential thermal analysis is performed using the instrument for synchronous thermal analysis Setaram LabSys Evo. It is established that the melting temperature of produced high-entropy alloy equals to 1495.18 °C.

In order to reveal the regularities of structure formation the samples of three regions locating at different distances from the substrate (15, 35, and 55 mm) were prepared. The samples were studied by the X-ray phase analysis method. Comparison of the obtained X-ray diffraction patterns (Fig. 6.9) detected no shift of diffraction peaks, as well as no change in the value of the ratio of diffraction peaks intensity to maximum peak intensity, was found. The obtained results are indicative of phase composition homogeneity of manufactured high-entropy alloy.

Figure 6.10 represents the results of diffractogram pattern indexing of HEA under study. It is determined that the sample crystal lattice has a cubic system. The revealed diffraction maximums, in accordance with the sample represented in the research (Chumak et al. 2008), may be described in the frameworks of one crystal lattice, namely, $Fe_{0,258}Ni_{1,164}Al_{0,578}$. Chemical Formula: $Fe_{0.54}Ni_{0.836}Al_{0.624}$. Empirical Formula: $Al_{0.624}Fe_{0.54}Ni_{0.836}$. Weight %: $Al_{17.53}Fe_{31.39}Ni_{51.08}$. Atomic %: $Al_{31.20}Fe_{27.00}Ni_{41.80}$. Compound Name: Aluminium Iron Nickel. Crystal (Symmetry Allowed): Centrosymmetric. AC Space Group: Pm-3 m (221). Pearson Symbol:

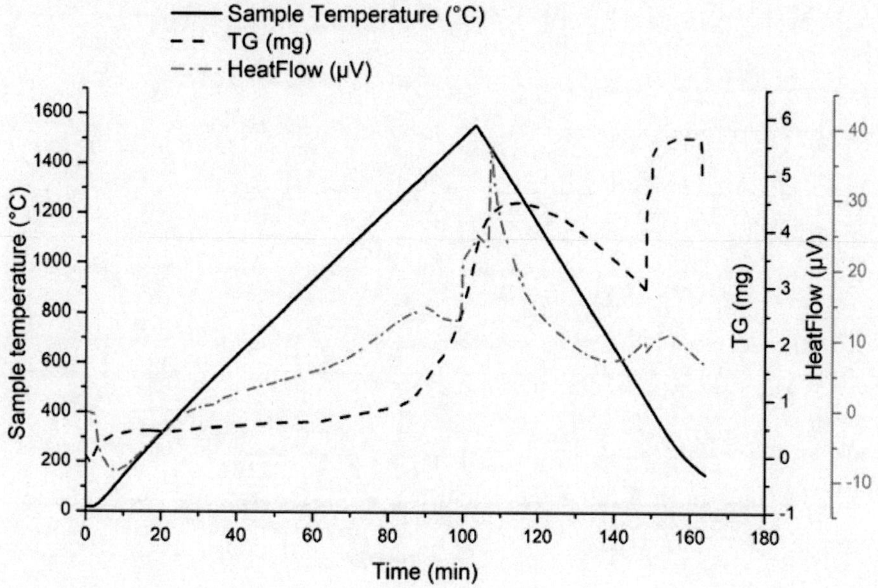

Fig. 6.8 Thermogravimetric analysis of HEA in its initial state

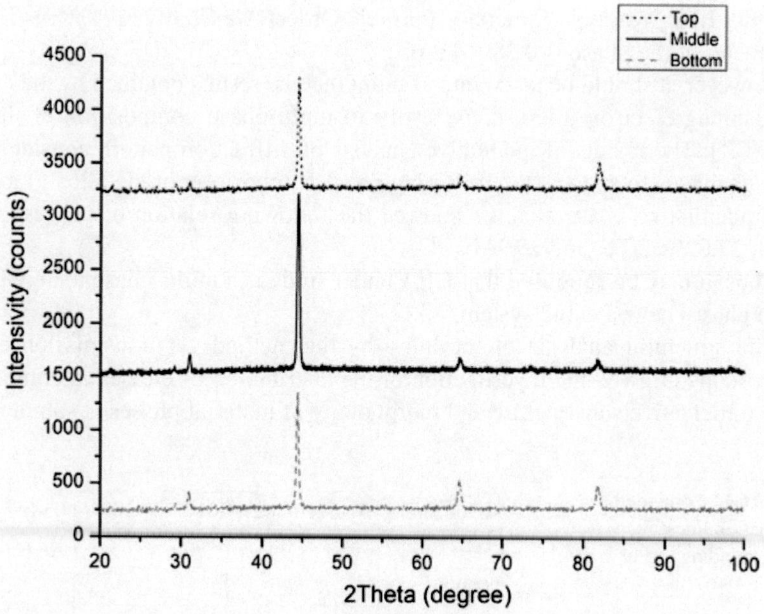

Fig. 6.9 X-ray diffraction patterns obtained in different regions of the HEA sample (top, middle and bottom)

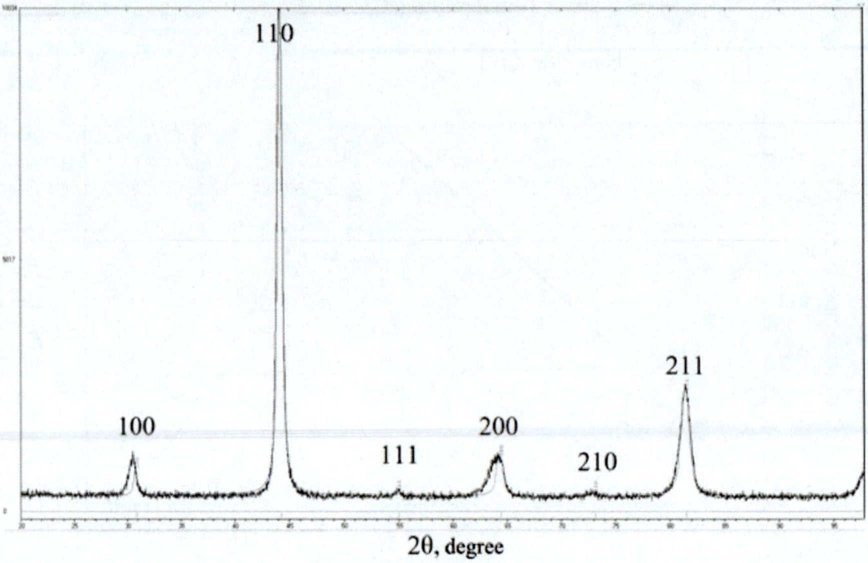

Fig. 6.10 X-ray diffraction pattern portion obtained from HEA sample

cP2. 00. LPF Prototype Structure [Formula Order]: Cs Cl, cf2, 221. Crystal lattice parameter of the phase is 0.28914 nm.

However, it should be taken into account that the results obtained by the methods of scanning electron microscopy testify to multi-phase composition of the alloy (Fig. 6.4). The results of qualitative analysis of diffraction pattern considering the facts enable us to suggest the following phase composition of the alloy (Table 6.1). The quantitative phase analysis showed the following relation of phases: 36.56% AlNi, 27.02% CrFe, 36.42% Al_2FeCo.

Thus, it may be supposed that HEA under study is a multi-phase material whose main phases have a cubic system.

The structural analysis of the alloy by the methods of transmission electron microscopy allowed the investigation of the distribution of elemental composition, state of defective substructure and morphology of material phases at submicro- and

Table 6.1 Coincided diffraction lines corresponding to Fig. 6.10

No. of card	01-071-5882	01-077-7598	03-065-4920
Phase	AlNi	CrFe	Al_2FeCo
3.06000			
2.88300	2.882		2.883
2.03950	2.03788	2.03505	2.03859
1.44180	1.441	1.439	1.4415
1.17730	1.17657		1.17698
1.01930	1.01894		1.01929

nanodimensional levels. It is detected that inclusions of two-dimensional classes: submicron inclusions (Fig. 6.11a–c) and nanodimensional inclusions (Fig. 6.11d, particles are indicated by arrows) are present in the alloy.

The investigation into the elemental composition of the particles revealed in the alloy is performed by the methods of micro-X-ray spectral analysis (Fig. 6.12). The results of the quantitative analysis of particles' elemental composition are presented in Table 6.2.

When analysing the results presented in Table 6.2 the presence of some amount of oxygen and carbon atoms can be noted in the analysed portion of foil, along with the specified elements (Al, Fe, Cr, Co, Ni). In this case, the carbon atoms are detected mainly in the bulk of second phase inclusions. Therefore, on the basis of the results of micro-X-ray spectral analysis, it may be supposed that the inclusions being present on the foil portion are carbides. The main elements of carbides are chromium and iron.

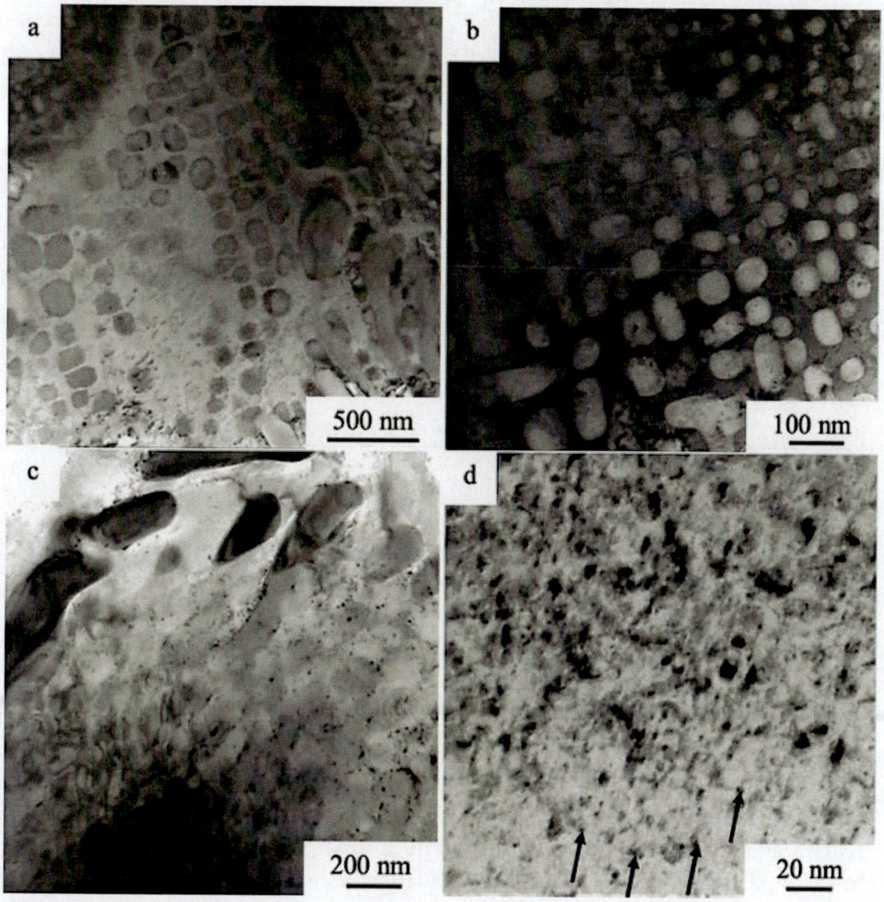

Fig. 6.11 Electron microscopic images of the HEA structure obtained studying thin foils

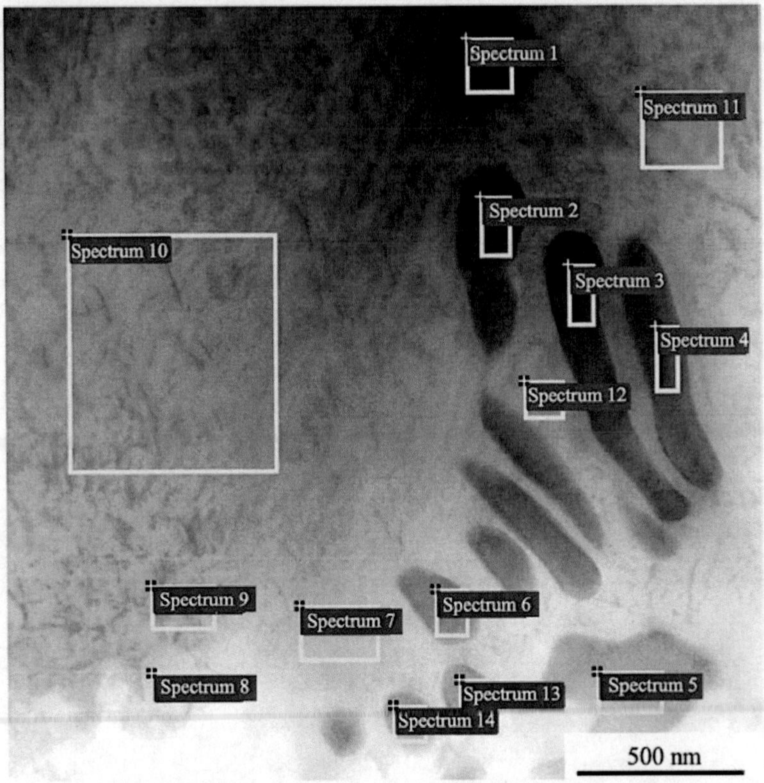

Fig. 6.12 STEM image of the HEA foil section. The areas from which micro-X-ray analysis of the alloy was done are indicated

6.3 Mechanical Properties of AlCoCrFeNi High-Entropy Alloy

The HEA mechanical properties were characterized by microhardness value being determined on microhardness tester HV-1000 and nanohardness value (nanotester NanoScan-4D). Measurements were taken both along deposited layers and in a perpendicular, cross-section to detect the degrees of material homogeneity.

The microhardness analysis by Vickers method at indenter load of 1 N was carried out and it was performed using 5 or more measurements in 14 areas in every 5 mm of sample cross-section. The 4% increase in microhardness value was found in the area located on boundary with the substrate that may be related to the 10% increase in iron content revealed in the course of layer-by-layer elemental analysis. In the remaining areas, the value deviates from the average one by no more than 2%. As a result of performed studies it is established that the mean microhardness value of the samples produced by mode No. 1 amounts to 474 ± 18 HV, by mode No. 2—465 ± 3 HV, by mode No. 3—474 ± 8 HV.

Table 6.2 The results of the quantitative analysis of the elemental composition of foil portion in atomic%. Its STEM image is shown in Fig. 6.12

Spectrum	C	O	Al	Si	Cr	Fe	Co	Ni
Spectrum 1	46.60	1.07	0.88	0.44	40.38	6.59	1.00	3.05
Spectrum 2	52.95	−0.08	1.02	0.49	35.55	6.09	0.94	3.04
Spectrum 3	50.71	1.60	0.88	0.66	35.29	6.51	0.88	3.49
Spectrum 4	53.50	1.28	1.47	0.45	31.82	6.52	1.01	3.95
Spectrum 5	64.44	3.49	0.85	0.57	19.47	4.89	1.05	5.23
Spectrum 6	63.23	2.53	1.39	0.80	20.82	5.49	1.07	4.67
Spectrum 7	30.80	5.85	9.95	3.03	12.62	19.83	3.88	14.03
Spectrum 8	59.41	5.16	9.61	1.83	2.43	6.12	2.13	13.31
Spectrum 9	37.40	1.55	12.27	2.65	10.50	16.76	3.52	15.35
Spectrum 10	13.03	5.25	7.44	4.27	21.99	30.76	4.92	12.35
Spectrum 11	20.44	2.21	19.99	2.07	10.55	18.65	4.15	21.94
Spectrum 12	37.35	1.21	10.61	3.01	12.87	18.61	3.44	12.90
Spectrum 13	66.20	−0.43	1.57	0.86	19.37	5.46	1.34	5.64
Spectrum 14	63.67	1.91	1.31	0.91	18.94	6.32	1.39	5.56

The measurement of Rockwell hardness by indentation of the diamond tip with a load of 150 kgf depending on distance from substrate showed a growth of hardness value also in the area adjacent to the substrate. The hardness amounted to 9% compared to the mean value. The Rockwell hardness measured in 3 areas located at distances of 50, 30, and 10 mm from substrate amounted to 47 ± 5 HRC on average. The obtained hardness is in correlation with the hardness of steel grade 45 (45 HRC).

Averaged microhardness value of samples measured along the whole length of the cross-section in every 5 mm amounts to 463 ± 3 HV, which on the whole is indicative of its uniform distribution (Fig. 6.13).

On average, the nanohardness of sample under study amounts to 10.4 ± 0.8 GPa independent of distance to surface. The elastic modulus measured by nanoindentation method amounts to 304 ± 15 GPa that correlates, as will be shown below, with the data obtained under compression tests. Thus, based on the obtained values of mechanical properties of high-entropy alloy it may be concluded that it has a high hardness.

Tribological properties (wear resistance and friction factor) were determined on planar samples on tribometer Pin on Disc and Oscillating TRIBOtester (TRIBOtechnic, France) with the following parameters: 6 mm diameter ball made of Al_2O_3 ceramic material, the sample rotation rate of 25 mm/s, path travelled by counterbody of 100 m, indenter load of 5 N, wear track radius of 2 mm, at room temperature and normal humidity. Degree of material wear was determined by the results of track profilometry formed under tests. It is stated as a result of performed tests that HEA wear factor amounts to 1.4×10^{-4} $m^3/N{\cdot}m$, friction factor is 0.65.

Preliminary analysis of scientific sources showed that the tests for HEA plastic properties are carried out by the most frequent method of uniaxial compression

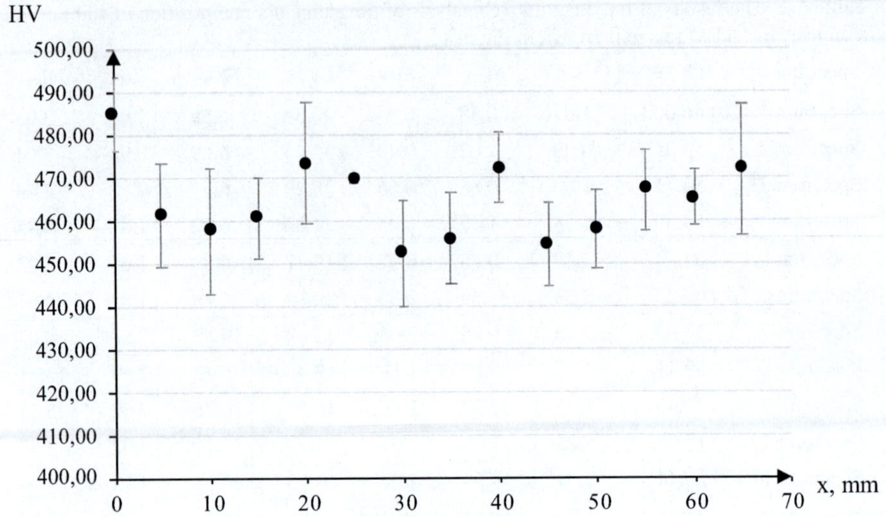

Fig. 6.13 Plot of HEA microhardness obtained by mode No. 2 as a function of distance from the substrate

(Zhang et al. 2016, 2018; Chen et al. 2015; Wang et al. 2009). In this context, a decision was made to perform the main tests by the method of uniaxial compression and the additional tests—by the method of uniaxial tension. To realize the tests by the method of uniaxial compression the cylindrical samples 10 mm in height and 5 mm in diameter were cut out of a HEA bulk sample by electroerosive method. For uniaxial tension tests the samples 2.3 mm thick × 9.1 mm wide × were 16.0 mm long (length of working part) were manufactured). For each of the three formation modes of Al-Co-Cr-Fe-Ni-system high-entropy alloy, no less than five cylindrical samples and five planar samples were fabricated. Tests for compression and tension were performed at room temperature.

A typical view of HEA samples prior to and after tests by the method of uniaxial compression is shown in Fig. 6.14.

Characteristic deformation curve obtained under compression of alloy samples manufactured by mode No. 2 is presented in Fig. 6.15.

The results of compression tests of HEA samples obtained at different modes of sample formation showed that the ultimate strength of material manufactured by mode No. 1 is in the interval from 652 to 856 MPa, by mode No. 2 it is in the interval from 1361 to 1899 MPa, by mode No. 3—it is in the interval from 556 to 1390 MPa. The Young's modulus of samples manufactured by the modes has the value of 165–212 GPa (mode No. 1), 273–372 GPa (mode No. 2), 192–270 GPa (mode No. 3). Therefore, the samples manufactured by mode No. 2 possess the largest value of conditional ultimate strength and Young's modulus compared to the samples manufactured by two other modes. It should also be noted that at the failure of samples manufactured by modes No. 1 and No. 3 the jumps on deformation

Fig. 6.14 **a** Typical sample view prior to compression, **b**—samples after compression tests

Fig. 6.15 Stress–strain curve of sample tested under uniaxial compression. Compression rate is 0.2 mm/min (3, 6 MPa/s)

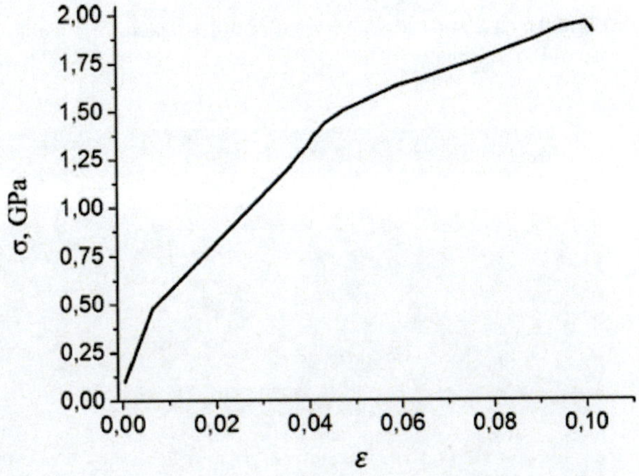

curves were observed. The latter may be caused by the presence of microcracks and micropores in samples manufactured by the modes (No. 1 and No. 3).

The results of tests for tension at temperature of 20 °C revealed a brittle failure of obtained samples independent of the place out of which the samples were cut from a solid massive sample. A typical curve of HEA stress–strain curve is presented in Fig. 6.16.

The failure occurred by the mechanism of intergranular cleavage along the plane located at an angle of 45° to the tension axis (Fig. 6.16b). Ultimate tensile strength amounted to 4 MPa on the average.

Scanning electron microscopy of failure surface showed that in the bulk of grains a grooved pattern was present, the pattern is stepped between different local cleavage facets of the same common plane (Fig. 6.17).

A source of local failure was the second phase particle located in the junction of four grains (the site of the particle location is indicated by the arrow in Fig. 6.18). Microrelief of adjacent grains is different that is indicative of their different crystallographic orientation. In Fig. 6.18b the arrow indicates a system of widely opened

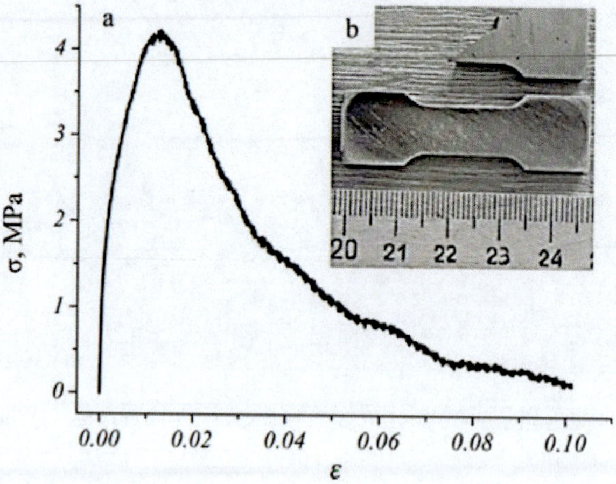

Fig. 6.16 **a** Stress–strain curve of samples tested for tension, **b** typical view of sample prior to and after tests for tension

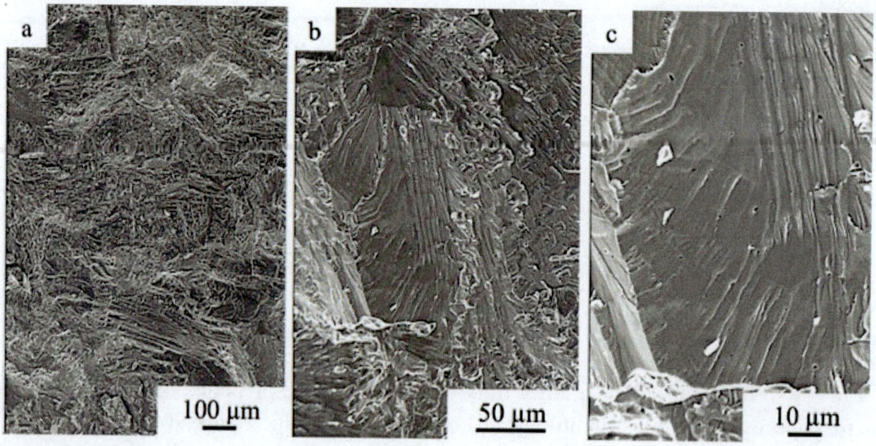

Fig. 6.17 SEM of HEA failure surface subjected to stretching deformation

secondary cracks that testifies the material cracking in the direction perpendicular to the main plane of failure.

Analysis of scientific sources containing data on mechanical properties of Al-Co-Cr-Fe-Ni-system high-entropy alloys manufactured by conventional technologies (Table 6.3) showed that ultimate strength of alloys varies (depending on the content of elements) from 370 to 2976 MPa, microhardness—from 100 to 637 HV, Young's modulus—from 187 to 251 GPa and nanohardness from 1.83 to 10.14 GPa.

Thus, Al-Co-Cr-Fe-Ni alloy produced in the research by layer-by-layer application on steel substrate by the technology of wire-arc additive manufacturing (WAAM) (cold metal transfer (CMT)) in the atmosphere of inert gas (Ar $\approx$ 99.99%) possess

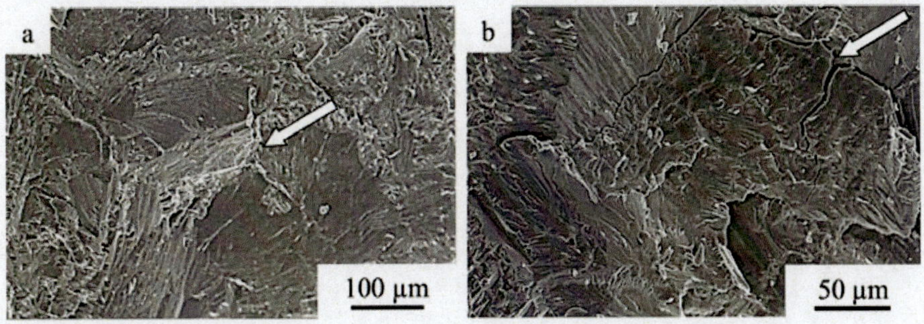

Fig. 6.18 SEM of characteristic elements of HEA failure surface structure subjected to tension deformation. The particle of the second phase being the source of sample local failure is indicated by the arrow

the properties comparable with properties of alloys being produced by conventional technologies listed in Table 6.3.

6.4 Structure, Phase Composition and Properties of HEA Alloys Irradiated with Pulsed Electron Beam

Irradiation of samples with intense pulsed electron beam was done on setup "SOLO" (Ivanov et al. 2009). Irradiation of samples was performed at the Institute of High-Current Electronics SB RAS. The samples fabricated for uniaxial compression tests (cylindrical samples) and planar samples to investigate into phase composition, defective substructure and to determine microhardness, nanohardness, wear parameter, friction factors, Young's modulus, et al. Electron-beam processing was performed with the following parameters: energy of accelerated electrons $U = 18$ keV, energy density of electron beam $E_s = (10, 15, 20, 25, 30)$ J/cm^2, pulse duration of electron beam $f = 200$ μm, pulse number $N = 3$. Irradiation was done in vacuum at a pressure of residual gas (argon) in setup chamber $p = 0.02$ Pa.

The series of Figs. 6.19, 6.20 and 6.21 represents images of HEA surface structure being formed on irradiation with pulsed electron beam at different energy density of electron beam. When analysing the results shown in the figures the following features of forming the structure of irradiation surface may be noted. First, we emphasize that independent of the energy density of electron beam the HEA irradiation is accompanied by fragmentation of sample surface by microcracks' network. The fragments sizes reach several hundred micrometres, essentially increasing in grain sizes of initial alloy. It is evident that microcracks form as a result of relaxations of elastic stresses arising in material surface layer on velocity cooling taking place in conditions of irradiation by pulsed electron beam of submillisecond duration effect. The relaxations of elastic stresses by microcrack formation are typical of ceramic materials and are indicative of increased brittleness of HEA under study.

Table 6.3 Values of mechanical properties of Al-Co-Cr-Fe-Ni-system alloys manufactured by conventional technologies

Alloy	Microhardness (HV)	Nanohardness (GPa)	Elastic modulus (GPa)	Yield strength (MPa)	Ultimate tensile strength (MPa)	Ultimate compression strength (MPa)
AlCoCrFeNi	560 (Jiao et al. 2015)	10.14 (Tian et al. 2019)	251.45 (Tian et al. 2019)	816 (Alagarsamy et al. 2016)	2835 (Alagarsamy et al. 2016)	2003 (Alagarsamy et al. 2016)
CoCrFeNi	100 (Kang et al. 2018)					
Al0.1CoCrFeNi	100 (Kang et al. 2018)	1.83 ± 0.12 (Guo et al. 2017)	203 ± 17 (Guo et al. 2017)	140 ± 6 (Sun et al. 2019)		370 ± 8 (Sun et al. 2019)
Al0.3CoCrFeNi	105 (Kang et al. 2018)					
Al0.4CoCrFeNi	105 (Kang et al. 2018)					
Al0.5CoCrFeNi	200 (Kang et al. 2018)					
Al0.7CoCrFeNi	350 (Kang et al. 2018)	4.57 ± 0.09 (Guo et al. 2017)	235 ± 11 (Guo et al. 2017)			
Al0.8CoCrFeNi	400 (Kang et al. 2018)					
Al0.9CoCrFeNi	525 (Kang et al. 2018)					
AlCoCrFeNi	510 (Kang et al. 2018)					
Al1.2CoCrFeNi	490 (Kang et al. 2018)					

(continued)

Table 6.3 (continued)

Alloy	Microhardness (HV)	Nanohardness (GPa)	Elastic modulus (GPa)	Yield strength (MPa)	Ultimate tensile strength (MPa)	Ultimate compression strength (MPa)
Al1.5CoCrFeNi	490 (Kang et al. 2018)					
Al1.8CoCrFeNi	490 (Kang et al. 2018)					
Al2.0CoCrFeNi	500 (Kang et al. 2018)					
AlCrFeNi				1271,2 (Chen et al. 2015)	1930,8 (Chen et al. 2015)	
AlCo0.25CrFeNi				1363,2 (Chen et al. 2015)	2161,1 (Chen et al. 2015)	
AlCo0.5CrFeNi				1349 (Chen et al. 2015)	2047,6 (Chen et al. 2015)	
AlCo0.75CrFeNi				1331,4 (Chen et al. 2015)	1973,2 (Chen et al. 2015)	
AlCoCrFeNi				313,7 (Chen et al. 2015)	1943,2 (Chen et al. 2015)	
AlCoCr0.3FeNi	415 (Qin et al. 2018)			1109 (Qin et al. 2018)		1579 (Qin et al. 2018)

(continued)

Table 6.3 (continued)

Alloy	Microhardness (HV)	Nanohardness (GPa)	Elastic modulus (GPa)	Yield strength (MPa)	Ultimate tensile strength (MPa)	Ultimate compression strength (MPa)
AlCoCr0.5FeNi	431 (Qin et al. 2018)			1144 (Qin et al. 2018)		1759 (Qin et al. 2018)
AlCoCr0.7FeNi	448 (Qin et al. 2018)			1267 (Qin et al. 2018)		1826 (Qin et al. 2018)
AlCoCrFeNi	498 (Qin et al. 2018)			1394 (Qin et al. 2018)		1841 (Qin et al. 2018)
AlCoCrFe0.2Ni	637,2 (Cao et al. 2019)					2335 (Cao et al. 2019)
AlCoCrFe0.4Ni	605 (Cao et al. 2019)					2335 (Cao et al. 2019)
AlCoCrFe0.6Ni	542 (Cao et al. 2019)					2335 (Cao et al. 2019)
AlCoCrFe0.8Ni	502 (Cao et al. 2019)					2335 (Cao et al. 2019)
AlCoCrFe1.2Ni	487 (Cao et al. 2019)					2100 (Cao et al. 2019)

(continued)

Table 6.3 (continued)

Alloy	Microhardness (HV)	Nanohardness (GPa)	Elastic modulus (GPa)	Yield strength (MPa)	Ultimate tensile strength (MPa)	Ultimate compression strength (MPa)
AlCoCrFe1.4Ni	483 (Cao et al. 2019)					2100 (Cao et al. 2019)
AlCoCrFe1.6Ni	480 (Cao et al. 2019)					2100 (Cao et al. 2019)
AlCoCrFe1.8Ni	477 (Cao et al. 2019)					2100 (Cao et al. 2019)
AlCoCrFe2.0Ni	460,2 (Cao et al. 2019)					2100 (Cao et al. 2019)
AlCoCrFeNi	488 (Takeuchi and Inoue 2016)			1350 (Takeuchi and Inoue 2016)	2840 (Takeuchi and Inoue 2016)	
AlCoCrFeNi1.5	524 (Takeuchi and Inoue 2016)			1360 (Takeuchi and Inoue 2016)	2 610 (Takeuchi and Inoue 2016)	
AlCoCrFeNi1.8	330 (Takeuchi and Inoue 2016)			600 (Takeuchi and Inoue 2016)	2470 (Takeuchi and Inoue 2016)	
AlCoCrFeNi2.1	300 (Takeuchi and Inoue 2016)			670 (Takeuchi and Inoue 2016)	–	
AlCoCrFeNi2.4	270 (Takeuchi and Inoue 2016)			500 (Takeuchi and Inoue 2016)	2430 (Takeuchi and Inoue 2016)	
AlCoCrFeNi2.7	260 (Takeuchi and Inoue 2016)			470 (Takeuchi and Inoue 2016)	2790 (Takeuchi and Inoue 2016)	
AlCoCrFeNi3.0	241 (Takeuchi and Inoue 2016)			440 (Takeuchi and Inoue 2016)	2380 (Takeuchi and Inoue 2016)	

(continued)

Table 6.3 (continued)

Alloy	Microhardness (HV)	Nanohardness (GPa)	Elastic modulus (GPa)	Yield strength (MPa)	Ultimate tensile strength (MPa)	Ultimate compression strength (MPa)
Al0.7NiCoFe1.5Cr1.5	562 (Ivanov et al. 2009)			1523 (Ivanov et al. 2009)		
Al0.25Co1.75CrFeNi	138 (Ivanov et al. 2009)					
Al0.5Co1.5CrFeNi	325 (Ivanov et al. 2009)					
Al0.75Co1.25CrFeNi	328 (Ivanov et al. 2009)					
Al1.25Co0.75CrFeNi	510 (Ivanov et al. 2009)					
Al1.5Co0.5CrFeNi	540 (Ivanov et al. 2009)					
Al1.75Co0.25CrFeNi	603 (Ivanov et al. 2009)					
AlCrFeNi				1300 (Qin et al. 2018)		2743 (Qin et al. 2018)
(AlCoCrFeNi)96Co4				1350 (Qin et al. 2018)		2976 (Qin et al. 2018)
(AlCoCrFeNi)92Co8				1300 (Qin et al. 2018)		2784 (Qin et al. 2018)
(AlCoCrFeNi)88Co12				1080 (Qin et al. 2018)		2380 (Qin et al. 2018)
(AlCoCrFeNi)84Co16				750 (Qin et al. 2018)		2680 (Qin et al. 2018)

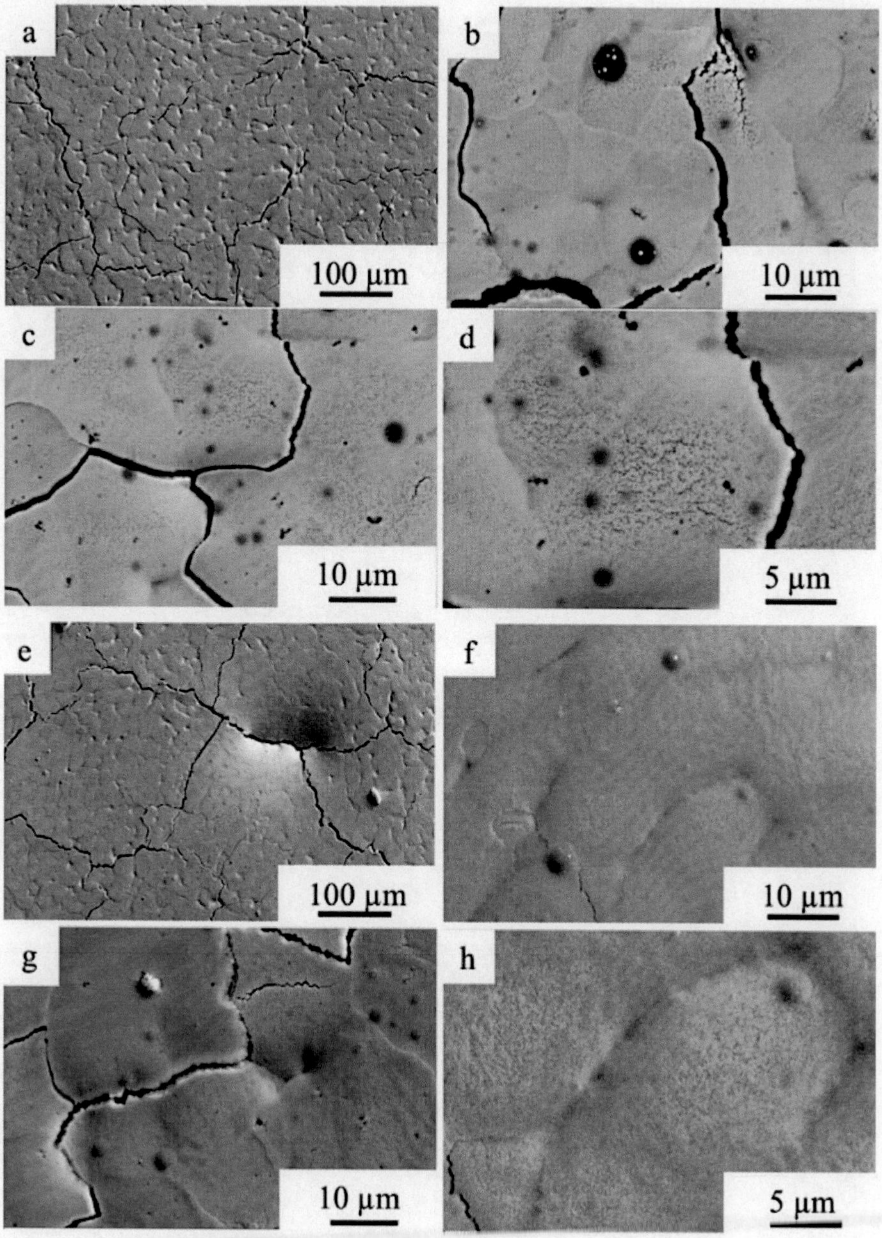

Fig. 6.19 SEM image of HEA structure being formed on irradiation by pulsed electron beam (E_S = 10 J/cm^2 (**a-d**) and 15 J/cm^2 (**e–h**))

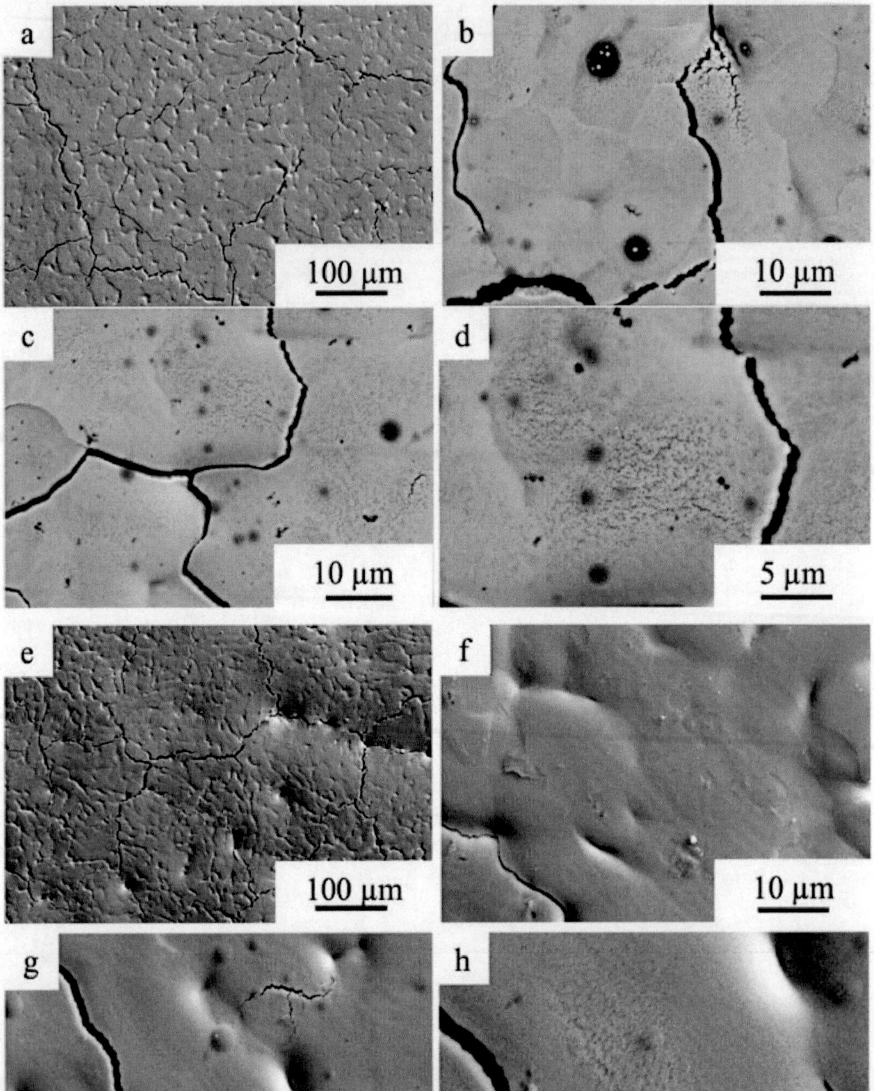

Fig. 6.20 SEM image of the HEA structure being formed on irradiation by pulsed electron beam ($E_S = 20$ J/cm^2 (**a**-**d**) and 25 J/cm^2 (**e**–**h**))

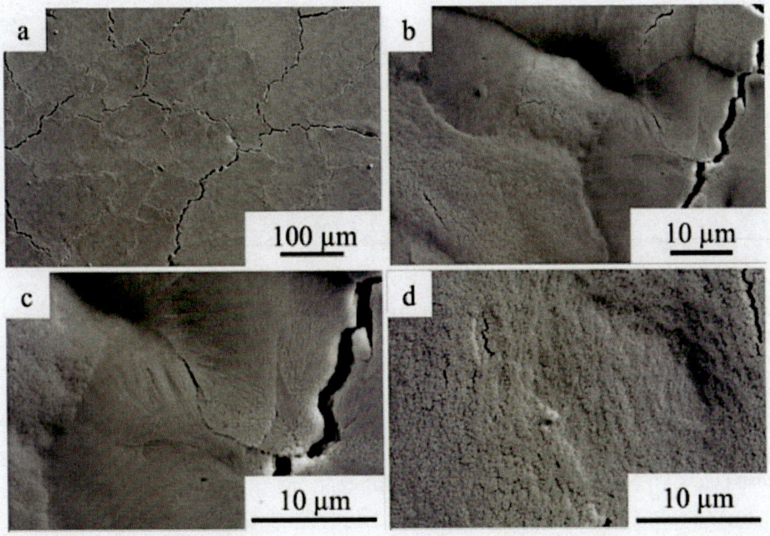

Fig. 6.21 SEM image of the HEA structure being formed on irradiation by pulsed electron beam ($E_S = 30$ J/cm^2)

The HEA irradiation by the pulsed electron beam is the mode of surface layer melting is accompanied by material homogenisation as evidenced by the release of grain boundaries from second phase precipitations. The increase in energy density of electron beam results in intensification of the formation process of alloy homogeneous in elemental composition.

High-velocity crystallisation of melted surface layer of HEA samples is accompanied by the formation of submicronanocrystalline structure (Fig. 6.22).

The size of crystallites increases with the growth of energy density of electron beam and vary at $E_S = 30$ J/cm^2 in the limits from 100 to 200 nm (Fig. 6.22b, c).

There is some probability that alloy irradiation by pulsed electron beam in mode of surface layer melting may result in uncontrolled change in elemental composition

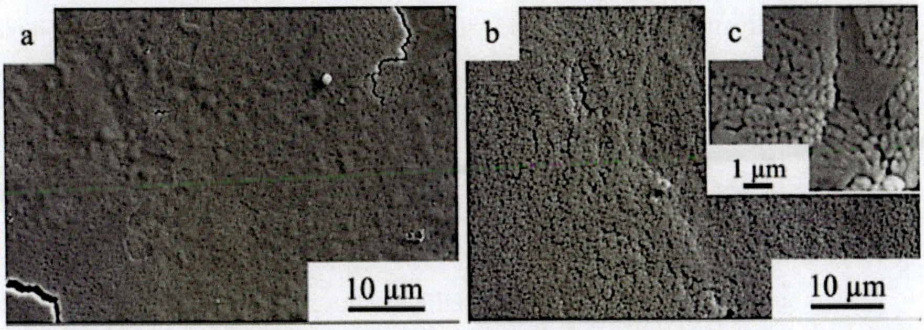

Fig. 6.22 Electron microscopic image of HEA surface structure irradiated by pulsed electron beam (200 μm, 3 pulses) at energy density of electron beam 30 J/cm^2

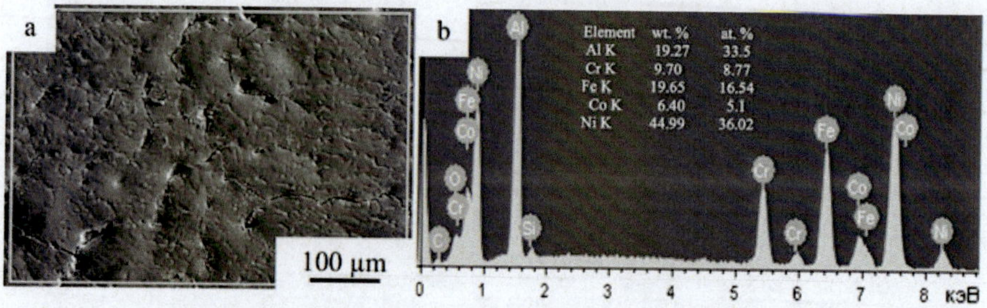

Fig. 6.23 Electron microscopic image **a** of HEA surface portion irradiated by pulsed electron beam (10 J/cm^2, 200 μs, 3 pulses); **b**—energy spectra obtained from the section (**a**). The table lists the results of the micro-X-ray spectral analysis

Table 6.4 Results of micro-X-ray spectral analysis of HEA samples in initial state and irradiated by pulsed electron beam (200 μs, 3 pulses) at different energy density of electron beam

E_S, J/cm^2	Concentration, at. %				
	Al	Cr	Fe	Co	Ni
Initial state	36.5	8.6	16.4	4.9	33.7
10	33.1	8.8	16.8	5.0	36.3
15	34.4	8.2	16.0	5.0	36.4
20	34.8	7.4	15.3	5.3	37.2
25	34.1	8.3	16.4	5.0	36.2
30	32.8	8.0	16.0	5.3	37.9

of the material. The investigations into elemental composition of HEA surface layer irradiated by pulsed electron beam carried out by the methods of the micro-X-ray spectral analysis proved no truth of the apprehensions. In fact, as follows from the analysis of results shown in Fig. 6.23 and Table 6.4 the elemental composition of the alloy surface layer is practically independent of the energy density of the electron beam and corresponds to the elemental composition of initial material within the limits of measurement error.

The investigations into the surface structure of brittle fracture of HEA samples modified by pulsed electron beam were carried out by the methods of scanning electron microscopy. It is established that the high-velocity crystallisation of surface layer results in the formation of columnar structure whose characteristic image is shown in Fig. 6.24.

Thickness of modified layer (H) increases from 0.8 μm to 20 μm in a regular way with increase in energy density of electron beam from 10 J/cm^2 to 30 J/cm^2 (Table 6.5).

Columnar structure is formed by crystallites whose sizes (h) increase in a regular way with the growth of energy density of electron beam (Table 6.5). Thus, such parameter as energy density of electron beam affects both the thickness of modified layer and size of crystallites being formed as a result of processing.